Goldfish Varieties and Genetics

A Handbook for Breeders

Joseph Smartt

Illustrations by
Merlin Cunliffe

Fishing News Books
An imprint of Blackwell Science

Blackwell Science

Fishing News Books
A division of Blackwell Science Ltd
Editorial Offices:
Osney Mead, Oxford OX2 0EL
25 John Street, London WC1N 2BS
23 Ainslie Place, Edinburgh EH3 6AJ
350 Main Street, Malden
MA 02148 5018, USA
54 University Street, Carlton
Victoria 3053, Australia
10, rue Casimir Delavigne
75006 Paris, France

Other Editorial Offices:

Blackwell Wissenschafts-Verlag GmbH
Kurfürstendamm 57
10707 Berlin, Germany

Blackwell Science KK
MG Kodenmacho Building
7–10 Kodenmacho Nihombashi
Chuo-ku, Tokyo 104, Japan

Iowa State University Press
A Blackwell Science Company
2121 S. State Avenue
Ames, Iowa 50014–8300, USA

First published 2001

Set in 10/13 pt Times
by Sparks Computer Solutions Ltd, Oxford
http://www.sparks.co.uk

DISTRIBUTORS
Marston Book Services Ltd
PO Box 269
Abingdon
Oxon OX14 4YN
(*Orders:* Tel: 01865 206206
Fax: 01865 721205
Telex: 83355 MEDBOK G)

USA and Canada
Iowa State University Press
A Blackwell Science Company
2121 S. State Avenue
Ames, Iowa 50014-8300
(*Orders:* Tel: 800-862-6657
Fax: 515-292-3348
Web www.isupress.com
email: orders@isupress.com

Australia
Blackwell Science Pty Ltd
54 University Street
Carlton, Victoria 3053
(*Orders:* Tel: 03 9347 0300
Fax: 03 9347 5001)

A catalogue record for this title is available from the British Library

ISBN 0-85238-265-0

Library of Congress
Cataloging-in-Publication Data
Smartt, J.
Goldfish varieties and genetics: a handbook for breeders/Joseph Smartt; illustrations by Merlin Cunliffe.
p. cm.
Includes bibliographical references (p.).
ISBN 0-85238-265-0
1. Goldfish. 2. Goldfish – Varieties. 3. Goldfish – Genetics. I. Title.

SF458.G6 S63 2001
639.3'7484–dc21 2001025153

For further information on Fishing News Books, visit our website:
http://www.blacksci.co.uk/fnb/

FSC
www.fsc.org
MIX
Paper from responsible sources
FSC® C013604

Contents

Preface

It is customary in a preface for an author to attempt a justification for the work presented to the public. I see no reason to break with this longstanding and hallowed tradition. It has struck me in reviewing recent literature on the goldfish that varieties can be treated in a very cursory fashion. In the early part of the last century very much more extensive treatments were published. Much of this information is still relevant and nearly all of it is inaccessible at the present time, and it seemed to be both worthwhile and, I hope, useful to review critically the literature of the goldfish over the past century and beyond with an appropriate update.

In order to appreciate how the present situation with regard to goldfish varieties has arisen, it is sensible to develop the genealogical approach of Matsui and to attempt a reconstruction of its evolutionary history. To do this requires the application of such genetical knowledge as we have and, by dint of careful thought tempered by practical experience, produce a credible working hypothesis of the evolutionary history of our subject. It is unfortunate that the majority of successful goldfish breeders are not exactly at ease with the subject matter of genetics. Truth to tell, this also applies to many aspiring and professional biologists. What I have attempted to do is to present a highly selective summary of the relevant genetical principles. The reader might find it convenient to see, first of all, what they make of the chapter on goldfish variety evolution and refer back as necessary to the consideration of the relevant principles in the previous chapter.

At the present time among geneticists there is virtually no interest in a systematic exploration of the Mendelian genetics of the goldfish. We have to content ourselves with the substantial early work of Chen and Matsui between the wars for the basic information we need, supplemented by some more recent work in Japan by Kajishima published in the 1970s. We are helped by the past publication by Matsui of data which at the time could not be interpreted satisfactorily. This can be reviewed in the light of subsequent advances since the 1930s of our knowledge of genetical principles, tempered with the knowledge and experience of successful breeders.

I have been very materially assisted in my work by contact with amateur and professional goldfish breeders as well as scientists over the past two decades. It is a very great pleasure to acknowledge the interest and support of two national goldfish societies, the Goldfish Society of Great Britain (GSGB) and the Goldfish Society of America (GFSA). I am deeply indebted to my mentors in these societies, Mr James H. Bundell and Mr A.I. Thommu, respectively, for their constant and sustained interest in what I attempted to do. In addition the membership of both societies has also been very supportive. In my travels abroad I have been fortunate to meet a number of goldfish enthusiasts in China and Japan. It was a particular pleasure, through the good offices of Mr Neal Teitler, to meet Kajishima Sensei and to have some discussion with him. I am indebted to Mr Michael Stewart, formerly Curator of the Ocean Park Aquarium in Hong Kong, for arranging a visit to the goldfish farm of the Chang brothers, Mr Jackie Chan and Mr Louis Chan in China. There I had truly the goldfish experience of a lifetime. The visit to the Hong Kong Goldfish Pagoda was equally memorable.

No less interesting have been visits to fish farms in the United States, most notably the Maryland-based Hunting Creek Fisheries. As well as being a major producer of goldfish it has been involved in developing intriguing new varieties. So far I have been unable to visit West Coast enthusiasts on their home ground but it has been interesting to visit those on the East Coast.

It is appropriate to acknowledge the support of the School of Biological Sciences of the University of Southampton, which has provided hospitality for my fish stocks. I have also appreciated the friendly interest of staff and students in the Biology Departments in my work activities. The forbearance of the latter on hearing about goldfish in a course on evolution was also appreciated.

The major source of illustrations for this work has been the drawings and paintings of Mr Merlin Cunliffe, the quality of which speaks for itself. These have been supplemented by photographs of my own goldfish taken by Mr Barry Lockyer of the School of Biological Sciences, to whom I am indeed grateful, and from significant earlier publications. These include Wolf's *Goldfish Breeds*, Smith's *Japanese Goldsfish*, Hervey and Hems' *The Goldfish*, Orme's *Fancy Goldfish Culture*, two drawings by Mr James Bundell, and from Chen's 1925 paper. These are acknowledged in the appropriate places.

I am indebted to Mrs Pat Wood for efficiently producing the typed manuscript and to Mr Tom Fryer of Sparks Computer Solutions for his patience in the course of producing the book.

Chapter 1
Introduction

While familiarity with the goldfish has bred affection rather than contempt there is no doubting that it does rather tend to be taken for granted. Yet if one explores its antecedents it soon becomes apparent that it is in fact a very remarkable creature indeed. Although its first contact with the human race was as a source of food, this contact developed into its domestication not as a farmed animal but for ornamental and aesthetic purposes. This process of domestication and diversification coincides broadly in its time frame with the second millennium. Perhaps the goldfish could be proclaimed as the Millennium fish!

The initial impetus to domestication of animals and plants has been utilitarian. The first creature to be fully domesticated was the wolf, which evolved into the enormous range of forms which can be considered as making up *Canis familiaris*, the domestic dog, which has become the best friend of the human race and also performs a vast range of useful functions not only as a companion but as a guardian and an indispensable partner in hunting and herding. It was, however, with the domestication of plants and the evolution of agriculture and farming systems that the human race developed civilisations. These depended on the farming systems becoming so efficient that substantial proportions of the population did not have to produce their own food but could become artisans and craftsmen, initially, and subsequently artists, musicians, soldiers, priests, politicians, and so on. The more efficient agriculture became, the more scope there was for other than strictly practical considerations to operate in the diversification of plants and animals under domestication. It was only in the latter part of the second millennium that purely aesthetic considerations came to the fore, a comparatively early example of this being the notorious tulip mania in Holland which followed its introduction in the sixteenth century. At this time we should note that in China the development of fancy goldfish was already well under way, but more of this later.

Human evolution has two components, the evolution of *Homo sapiens* as an organism – the anthropological element – and the evolution of human culture which itself has two aspects, the inanimate and the animate or the inorganic and organic. It is the latter which concerns us, in which the human race has moulded the form of other organisms including our present subject, the goldfish, which can reasonably be considered a human cultural artefact and its producers as being in a very real sense creative artists, working not with wood, stone, canvas or other inert materials but with the very stuff of life itself.

The goldfish is a member of the Carp family, the Cyprinidae, which in terms of evolutionary diversity is a very successful group. The members are characterised by an absence of teeth in the jaw, possession of a protrusile upper jaw and development of teeth in the pharynx (throat). There are some 1700 recognised species and it is the most species-rich of all families of freshwater fishes. It is to be found naturally in much of the Eurasian land mass, Japan, most islands of the East Indies, North America and Africa. Its ecology is primarily freshwater but some species tolerate brackish water well and can even breed in saline conditions. The palaeontological evidence suggests that the group is ancient, traceable back to more than 40 million years before the present time. Its absence from South America and Australasia has been related to the concept of Continental Drift in that the family evolved after the break-up of Pangaea (the original land mass) had begun and that the South American and Australian tectonic plates had already separated from the main land body.

The Cyprinidae can be regarded as a relatively unspecialised if not primitive group, at least as far as basic morphology, anatomy and physiology are concerned. This is not to say that the basic Cyprinid form has not been capable of adaptive and evolutionary change. In fact the pattern of evolutionary change, even if it has involved minor rather than fundamental modification, has over the ages produced a pattern of variation which has caused taxonomists considerable difficulties in producing a scheme of classification which makes good evolutionary sense and which is practically useful. The goldfish itself has not been immune to the problems of classification and nomenclature. Linnaeus (1758) named the goldfish *Cyprinus auratus*, the name by which it was known to Darwin. The genus *Cyprinus* as defined by Linnaeus was much more extensive than it is at the present time and included all the British native cyprinids. Quite sensibly the genus has been split into several genera but it could be argued that this process has gone too far. In some texts every native British cyprinid has been placed in a different genus, which seems absurd. It is equally absurd that the carp is in a different genus from the goldfish and the Crucian carp. It almost appears to some taxonomists that if species can be distinguished they must be placed in different genera. The function of a good taxonomy and system of nomenclature is to convey both difference and relationship. This the classical binomial system does very well, but this purpose is defeated when every species finds itself placed in a monotypic genus. By contrast, the large genus *Barbus*, which included widely differing forms, defied division for a long time and even now opinions differ as to how the division should be made. However, the name *Carassius auratus* has been hallowed by usage of over a century and while one may protest, perforce one has to accept it. It is interesting that the goldfish specimen described by Linnaeus was a fancy twin-tailed individual and not the wild single-tailed progenitor type!

The carp family originated it is thought (Winfield & Nelson 1991) in East Asia with subsequent spread to the rest of Eurasia, Africa and North America. Both *Cyprinus* and *Carassius* followed this course of migration and these carps have made their way across Asia and into Europe. The origin of the goldfish in all its diversity has been in the very heartland of the whole family, which is in its way somehow appropriate. There has been a secondary dispersal of these carps throughout the world through human agency which has had interesting results. It has been questioned whether the carps are truly native to the western fringe of Eurasia. The forms of carp found in Britain suggest that, in the case of the common carp at least, there has been human assistance. The position is much more clear-cut in the case of North America, Africa, Australia and New Zealand where there were attempts by settlers from Europe to introduce and acclimatise useful species to their new homelands. (This resulted in the disastrous rabbit plague in Australia.) In retrospect, the successful attempt to introduce carp and goldfish to non-native areas has been a very mixed blessing. The carps are very hardy and adaptable, and feral populations of carp and goldfish have been established in North America, Africa and Australasia, where many consider they have had adverse effects on the indigenous fauna and its environment. As a result in many countries the introduction of exotic species capable of establishing themselves in local environments is now very strictly controlled or forbidden. In some instances this is a case of locking the stable door after the horse has bolted.

Feral populations of goldfish, particularly in the United States, have developed very interesting features. One of these can be mentioned briefly at this point. A mutant variant, called the 'bluebelly' in America, occurs at a relatively high frequency in the vicinity of Cleveland, Ohio, and has been studied over many years by Mr Al Thomma and collaborators. The bluebelly appears to be identical with the 'net-like transparent' mutant reported from the Far East and the 'mock-metallic' which was identified and studied by members of the Goldfish Society of Great Britain. Its high frequency in the Lake Erie area is probably due to the Founder Effect, which is often shown in large populations which have arisen from a small original parental group. This mutant has a reduced level of deposition of reflective guanine on the scales.

The position of the goldfish in the natural order can be conveyed best by the scheme of classification in Table 1.1.

In the nineteenth century the goldfish attracted some attention as a fitting object for scientific study. Darwin mentions it in his *Variation of Animals and Plants under Domestication* and over a century ago William Bateson considered it to be promising material for the scientific study of variation. His brief discussion of the goldfish and its variants has a surprisingly modern ring to it. It was in the Far East that the challenge was taken up most effectively in the twentieth century by Shisan Chen in China who published seminal work in the 1920s and 1950s and Yoshiichi Matsui in Japan whose most important work appeared in the 1930s. This work was mainly genetical and established a Mendelian basis for the genetics of the goldfish and the genetical control of the characters which were the distinctive features of fancy goldfish. However, although progress was undoubtedly achieved, a fully comprehensive understanding of goldfish genetics did not emerge. The reason for this was not apparent until the postwar period with the discovery that the goldfish had an unusual cytogenetic constitution. The chromosome counts of the goldfish and the common and Crucian carps were found to be double those of most other members of the Carp family. The commonest number of chromosomes in the Carp family (Cyprinidae) is 50, commonly presented as

Table 1.1 Classification of the goldfish.

Kingdom	Animalia
Phylum	Chordata
Sub-phylum	Vertebrata
Class	Osteichthyes
Sub-class	Actinopterygii
Infra-class	Teleostei
Super-order	Ostariophysi
Order	Cypriniformes
Family	Cyprinidae
Sub-family	Cyprininae
Genus	*Carassius*
Species	*auratus*

$2n = 50$. This is the somatic chromosome number, i.e. that found in most cells of the body. Half of the chromosomes are of paternal origin and half are maternal and come together when the egg is fertilised by the sperm. In the production of eggs and sperm, the chromosome numbers in both are reduced to the gametic number $n = 25$. The normal chromosome constitution is termed diploid for the body or soma and haploid for the gametes. The diploid chromosome complement is organised as 25 pairs of chromosomes which are termed homologous, that is they have a common ancestry and comparable genetic function. The genetic information in each of the pairs of homologues is similar (but not necessarily identical). The situation is greatly complicated when the chromosome number is doubled in polyploidy. Instead of each gene being present in duplicate it can be represented four times, which may convert a simple Mendelian monofactorial ratio 3 : 1 into the bifactorial 15 : 1. However there are added possibilities for complications which will be considered in more detail later. The added complexity of the genetic system has been a considerable constraint in achieving a broader understanding of what has been happening at the genetic level in response to the selection processes to which the goldfish has been subjected. The duplication of the chromosome complement (i.e. polyploidy) provides the organism with enhanced evolutionary potential in that additional genetic material can take on different genetic functions and increase the level of diversity which can develop.

Our understanding of the way in which the genetic system operates can be improved if we consider it in relation to the development of the goldfish as an embryo and subsequently. Several workers including Affleck (1952) have studied the embryology of the goldfish. It is during embryonic development that the development of the twin-tail phenotype (as seen in the Fantail) becomes apparent. Studies of the developmental biology of other traits such as telescope eyes, narial bouquets and hood growth have clearly demonstrated that these are post-embryonic developments. Apart from twin-tail development all other desirable mutant phenotypic characters develop after hatching. Mutations producing transparency of scales are expressed quite early in the fry-phase while the telescope and the celestial eye development takes place later. Hood development in Lionheads and Orandas occurs even later. Onset of colour change from the wild type to the characteristic red, orange or yellow is variable; it may be quite early at a few months or be delayed for a year and sometimes longer. In some colour variants, particularly the black melanic or the chocolate brown, there is no colour change as such. In the melanics there is progressive intensification of the black pigment, while the browns are distinguishable from all other metallic colour variants in showing a distinctive coloration right from the start, in the early development of the fry.

The reproductive biology of *Carassius* species is complicated by the occurrence of gynogenesis. This is related in some way to the occurrence of higher levels of polyploidy than the basic tetraploidy of the goldfish and common carp. Hexaploids $2n = 150$ and octoploids $2n = 200$ have been reported and studied in which gynogenesis is apparently the standard means of reproduction. In this process there is failure of the reduction division (meiosis) in which normally the number of chromosomes is reduced from the somatic ($2n$) level to the gametic (n). As a result, eggs are produced which contain the unreduced maternal somatic chromosome complement. In order to develop further these eggs with unreduced chromosome complements require activation. This is brought about by the penetration of the egg by a sperm, which takes no further part in the development of the egg; there is no fusion of the sperm and egg nuclei. Since all the genetic material of progeny comes from the mother, gynogenetic lineages are virtual clones, with the members of such gynogenetic clones all being females. Unfortunately complications do not end at this point: some tetraploid gynogenetic lines have been found and furthermore hexaploid males have also been identified. The latter apparently undergo normal meiosis in their spermatogenesis, and so, if there were females in which complete meiosis occurred in egg production, normal bisexual reproduction of hexaploids would become possible.

Commercially, gynogenesis has an undoubted attraction. If it were possible to induce outstanding individual goldfish to reproduce clonally by this means, it would improve efficiency of production enormously. It would remove the necessity for the sustained heavy culling necessary to maintain the quality of the best fancy goldfish lines. If the occurrence of gynogenesis could be controlled (initiated and reversed) then it would be possible to produce new improved forms by normal breeding and selection procedures and reproduce these selected individuals as clones. This is theoretically possible and could be carried out possibly even now on a small scale by appropriate biotechnological manipulation; whether it could be done on a large commercial scale economically is quite another matter.

Before we can consider the whole range of variants which have arisen in the course of domestication, we need to consider the form of the progenitor which produces this great array of diversity. The true carps are characterised by a long dorsal fin which covers approximately half the length of the dorsal contour from the snout to the tip of the caudal peduncle. The common carp is characterised by the possession of two pairs of barbels, which are absent in all members of the genus *Carassius*. The Crucian carp (*Carassius carassius*) has a characteristically deeper body than the goldfish, with 7–9 scales from the base of the dorsal spine to the lateral line, compared with 5–6 in the case of the goldfish. The number of scales along the length of the lateral line is greater in the Crucian carp than in the goldfish: 28–35 as against 25–30. This difference is partly related to the somewhat larger size of scales in the goldfish *vis-à-vis* the Crucian carp.

The form or morphology of the goldfish can be summarised as follows. Body form is relatively long and laterally compressed. The dorsal and ventral contours from the snout to the caudal peduncle are smooth. The body width is approximately two-thirds of its depth. Depth of body and length of the head are each about one-third of the body length. The lateral line describes a slightly sigmoid curve, forming an almost horizontal straight line. The height of the dorsal fin and the lengths of the pectoral, pelvic and anal fins are very similar and one-sixth of the body length generally. The dorsal and ventral lobes of the caudal fin are equal in length; the indentation of its trailing edge may be half the length of the lobes. Lobe length is around a quarter of the body length. It is important to bear these proportions in mind as they are very considerably changed in the varieties which have evolved under domestication and they serve as a yardstick by which the extent of such changes can be judged (Fig. 1.1).

Hervey and Hems (1948) have published a very full account of goldfish anatomy, to which the reader is referred. There are a number of very significant anatomical features which have become modified in the course of variety development, of which some note should now be taken in addition to the morphological aspects noted above. These concern the skeleton, which in many developed forms is shortened and the vertebral column becomes more strongly curved. This is reflected in the exaggerated curvature of the lateral line. The increase in depth of the body is a consequence of the decrease in length; growth in length is inhibited while growth in depth appears not to be (Fig. 1.2). This alteration in growth pattern produces

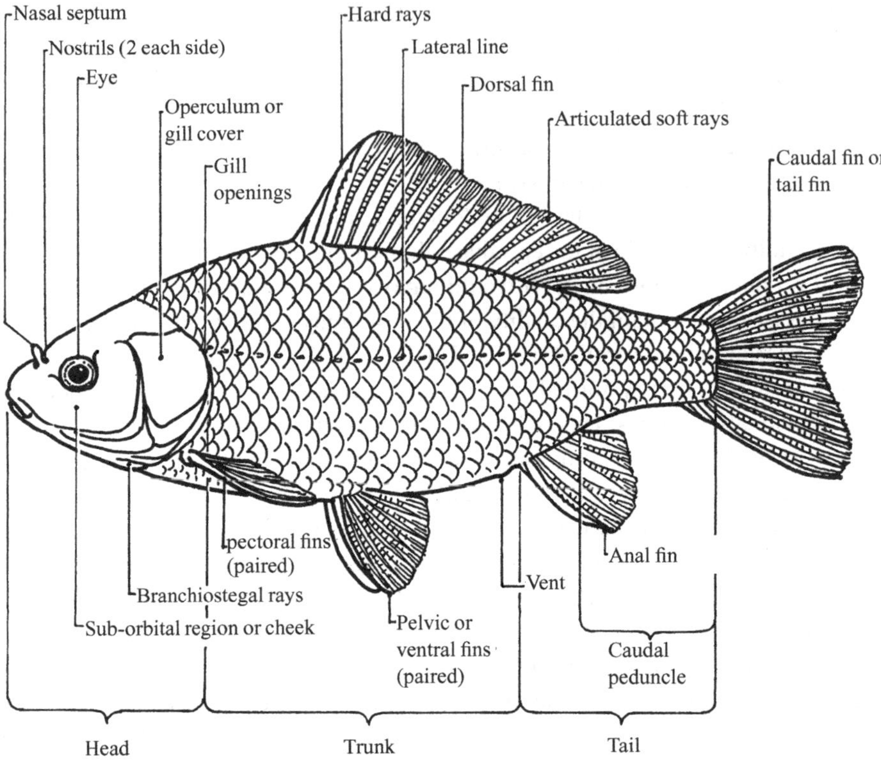

Fig. 1.1 Morphology of the Goldfish.

changes in the shape of the body cavity and the disposition of the organs it contains, principally the digestive tract and the swim bladder. The bulk of the internal organs appears to remain constant and in order to accommodate them in a shorter body cavity it has extended laterally, producing the characteristic and prized globular body form of the more highly evolved fancy goldfish (Fig. 1.3).

Perhaps the most remarkable feature developed in the course of goldfish evolution has been the twin-tail characteristic. In the development of the fin there are associated bony structures which are embedded in the muscles and are attached to the fin rays. Duplication of the caudal and anal fins is associated with the duplication of the supporting bony structures, in the case of the caudals the hypural bones and the interhaemal bones in that of the anals. Where development of fins is suppressed, for example the dorsal, the development of the supporting bones is also suppressed. If duplication of the supporting caudal fin bones is incomplete then imperfectly duplicated fins result. A similar situation applies to the dorsal fin when suppression of supporting bones is incomplete, development of dorsal spines and short fin segments may occur. Occasionally fish are seen in which the anal fin is completely absent; presumably during embryonic and larval development the supporting bony structure failed to develop and growth of the fin which normally occurs later is thereby inhibited. The full range of developmental possibilities which the occurrence of the twin-tail mutation sets in train concern

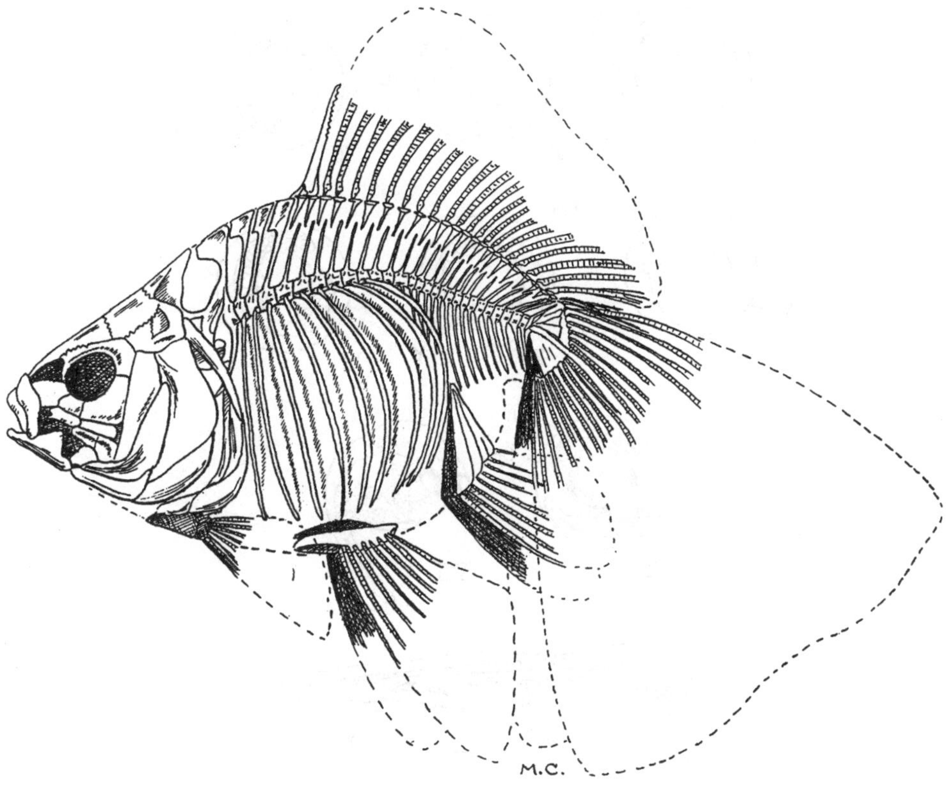

Fig. 1.2 Skeleton of the Veiltail goldfish.

Fig. 1.3 Relative proportions of swim-bladder lobes in: (a) Common Goldfish; (b) Comet; and (c) Veiltail (Hervey & Hems 1948).

not only the question of whether a single fin or duplicated fin develops but whether any fin develops at all (Figs 1.2, 1.4a–c). This question will receive more detailed attention later.

Undoubtedly the most striking difference between the ancestral wild goldfish and its domesticated descendants is in colour. That of the wild fish is cryptic and in nature is clearly advantageous, making the individual more difficult for predators to find. The domesticated goldfish generally is anything but inconspicuous and the bright coloration makes life easy for the heron and other predators on their visits to garden ponds. The whole population of such ponds may be wiped out in a single visit, whereas there is a sporting chance that in a wild-type population there might be survivors. Chances of survival in turbid waters for brightly coloured fish are higher and in feral populations brightly coloured fish can and do survive in such circumstances.

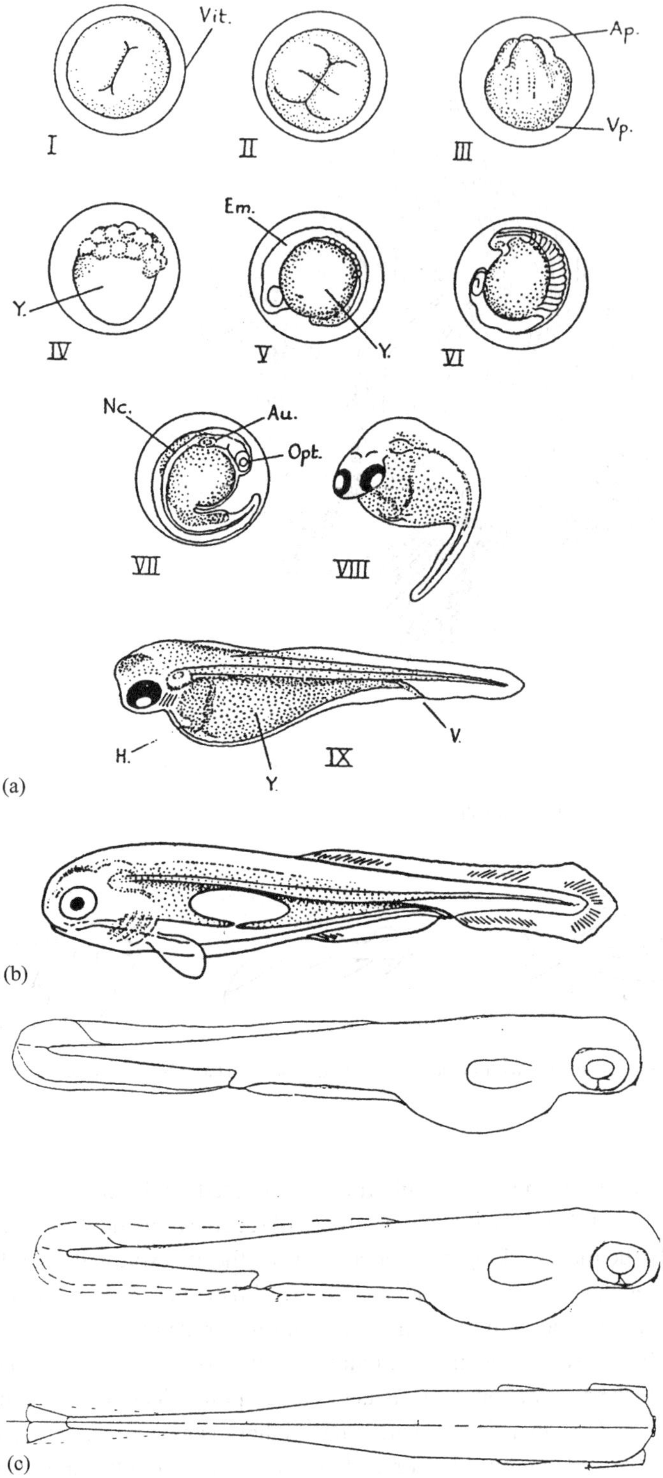

Fig. 1.4 (a) Embryology of the Goldfish (Hervey & Hems 1948). (b) Goldfish larva 80 hours post-hatching (Hervey & Hems 1948). (c) Morphology of twin-tailed goldfish larvae (J.H. Bundell).

In order to understand coloration of goldfish we need to appreciate how what we see is actually produced. In the goldfish there are three colour components, namely melanin which produces the darker components of colour, black, brown, blue and grey, the xanthic pigments which produce red, orange and yellow shades, and finally guanine which in the absence of the other two pigment types gives a white or silvery colour. Generally the effect of guanine is, when deposited on the scales, to give them a metallic sheen. If scales are transparent, guanine deposited in the dermis produces a mother-of-pearl or nacreous sheen. In the absence of melanin, xanthic pigments and guanine the body shows a pink tinge which is due to blood haemoglobin in superficial capillary blood vessels (Fig. 1.5).

Differences in colour can be produced by mutations which may act in a number of different ways: production of pigment may be entirely suppressed, the amount of pigment produced may be increased or decreased and the chemical nature of the pigment may be modified. The disposition of the pigment, especially in the case of melanin, is important. According to Orme (1979) there are four levels or depths at which pigment cells can occur and these can affect the colour that we actually perceive (Fig. 1.6). Eye colour in humans is produced by melanin; darker colours result from pigment cells located near the surface, deeper seated pigment produces hazel, green and blue eyes. In the complete absence of melanin, eyes are pink, as in true albinos. We see the same phenomenon in birds where coloration is strongly influenced by the presence or absence of melanin and its distribution. Black, brown, green and blue colorations are produced by melanin; the colours we perceive are produced by varied location of pigments in the feathers and the refractive properties of the feather matrix. The similar colour ranges we see in birds and the goldfish, as far as the melanin produced colours are concerned, arise in basically the same way.

In this relatively brief introductory chapter, I have attempted to introduce some of the important topics to be considered later. On the principle that we cannot understand the present or do anything to determine the future without an appreciation of the past, an account will be presented of the domestication and evolution of the goldfish. It is useful to understand how

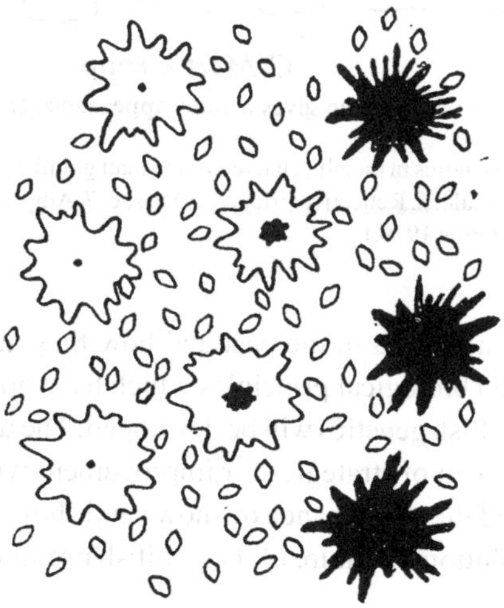

Fig. 1.5 Pigment cells (chromatophores). The pigment granules in those on the left are concentrated: in those on the right spread out (Hervey & Hems 1948).

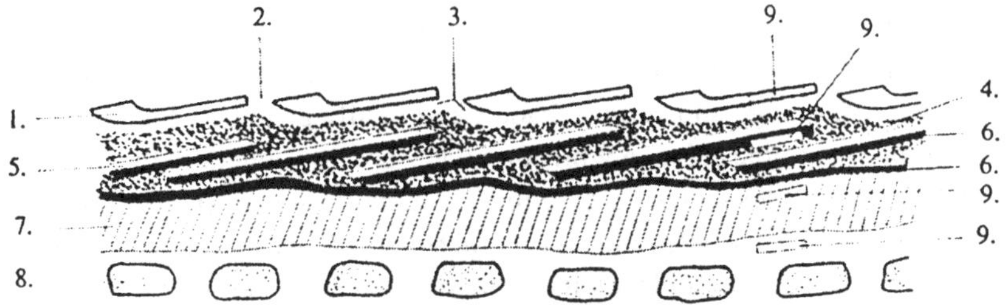

A ***Metallic Group***

Normal placement of reflective tissue below scales and deep in the dermis produces the well known metal-like shine of the common goldfish

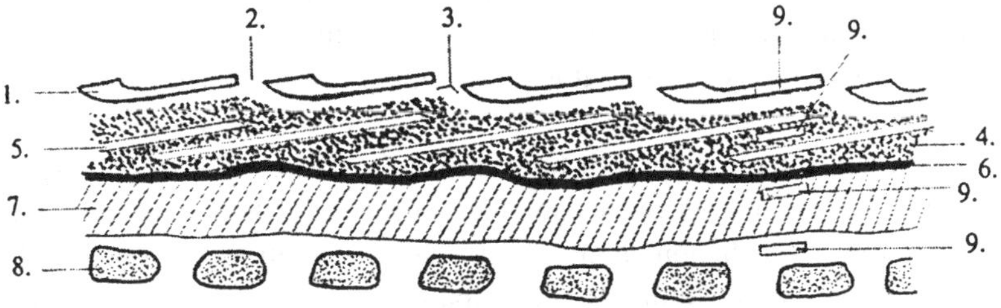

B ***Nacreous Group***

The upper layer of reflective tissue is missing, allowing the lower layer to show dully through the dermis and scales. This results in the fish having areas with a mother-of-pearl shine

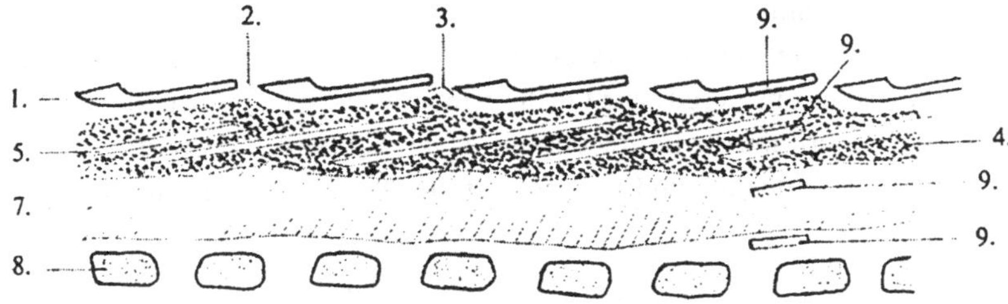

C ***Matt Group***

A complete absence of reflective tissue gives a matt appearance to the scales of the fish

Fig. 1.6 Distribution of chromatophores in metallic, nacreous and matt goldfish. 1. Epidermis. 2. Lateral line pore. 3. Lateral line canal. 4. Dermis. 5. Scale. 6. Reflective (iridoctyes) tissue. 7. Adipose tissue. 8. Myotomes – muscular tissue. 9. Chromatophore levels. (Orme 1979.)

goldfish varieties have been viewed in the past and how they are perceived at the present time. Relevant genetical and biological principles which have operated will be set out and a detailed presentation of goldfish genetics will be developed. The useful application of genetical principles and development of strategies for improvement will be outlined. The subject of goldfish appreciation and the significance of shows and show standards will be reviewed and an appraisal of the evolutionary future of the goldfish presented.

Chapter 2

History and Development of the Goldfish

In attempting to elucidate the history and development of the goldfish as we know it today, we are very much following in the footsteps of Charles Darwin. As Shisan Chen (1956) observed, 'In China we have not so many varieties of domesticated pigeon as Darwin observed by his investigations, but we do have one animal species of which not only is the degree of variation comparable to that of the domesticated pigeons of the world but its origin in the natural state is far easier to determine and the historical material as regards its varietal formation is even more complete. It is no other than the ubiquitous goldfish.'

It is highly significant that the very first chapter of Darwin's *Origin of Species* was devoted to variation under domestication, the study of which had supplied Darwin with the idea of selection being the driving force behind evolution. His interest in the subject resulted in a two-volume work, *The Variation of Animals and Plants under Domestication*. In the latter he specifically considers the goldfish, though only briefly. Whereas Darwin considered primarily morphological evidence, there were among his contemporaries some, most notably Alphonse de Candolle, a Swiss botanist who in his studies considered evidence from other, nonbiological, sources in elucidating the origins of cultivated plants. What is relevant for plants is also applicable to animals and this Chen has indicated in relation to the goldfish. The most significant feature of goldfish evolution under domestication is that it has all occurred in historical times in front of witnesses who were able to record both in writing and art works what they currently observed. While this record in the case of the goldfish is not as complete as we would like, it does put us in a very good position to produce a credible reconstruction of the series of events which produced the goldfish we know today. We are thus in some ways in a better position than those studying the domestication of domestic and farm animals, processes which go back 10 000–15 000 years, before civilisation developed.

Much of the goldfish literature considers the history of the goldfish but in general this tends to be anecdotal in nature. There are three noteworthy works which go back to the sources of information, that consider this in a rigorous scholarly fashion and give us a reliable and secure perspective on what has happened to the goldfish in the course of its domestication and what has transpired subsequently. These important works were all produced within a decade of the end of the Second World War. The first was *The Goldfish* produced in 1948 by Hervey and Hems, the second by Hervey in 1950 entitled *The Goldfish of China in the XVII Century*, and the third by Shisan Chen in 1956, to which reference has already been made. Hervey was very fortunate in securing the advice of Dr A. C. Moule, the eminent Cambridge sinologist whose contribution he gracefully acknowledges. These three works provide the substantial basis for our understanding of the domestication and evolution of the goldfish in the course of the second millennium.

Domestication was probably quite a long drawn-out process and it is logical to consider it in three stages or phases: pre-domestication, the domestication process and diversification under domestication. One can also ask a number of questions: when were goldfish domesticated, by whom, where this happened, how it came about and why. In attempting to find answers to these questions we may not achieve final answers but we shall certainly deepen our knowledge and understanding of the object of our interest and bring our overall perspective into sharp focus.

Pre-domestication

Successful domestication of both animals and plants occurred after the retreat of the ice sheets at the end of the last glaciation. The first animal to be domesticated was the wolf, an event which occurred probably within the last 15 000 years. This could well have been a complex process and initially may have been a rather casual business. *Homo sapiens* pre-dated the last glaciation and there is probably no good reason why some form of domestication or semi-domestication could not have been attempted earlier than actually transpired; if it was, it achieved no lasting success. Civilisation for its development depends on domestication primarily of plants and animals and was the outcome of the Neolithic Agricultural Revolution which had been successful at least 10 000 years ago in the Fertile Crescent of the Middle East. The motivation for such domestication was utilitarian, in order that human needs could be met more reliably in the first instance. After initial problems were overcome, agriculture became so successful that it became unnecessary for all the able-bodied to be engaged in subsistence food production. Labour surplus to meeting subsistence needs could be diverted to meet other needs, permitting development of crafts and the arts which have become one of the hallmarks of civilisation.

Domestication of any organism, plant or animal, is determined by the social environment in which it occurs. In China in the case of the goldfish there are three historical strands which are important: the social, the religious and the political. Once domestication is achieved political considerations tend to become more important. Initially the first important social consideration is the level of human cultural development which has been achieved. The full flowering of plant and animal domestication is seen in agricultural societies. The dog was domesticated in pre-agricultural times and was invaluable to hunter-gatherer communities. Its domestication facilitated that of herding livestock goats, sheep, cattle and pigs and less directly that of the horse. Initially pastoralists were also nomads and it is difficult to envisage domestication of fish by nomads. For this to come about, settled communities were necessary. Agriculture, it appears, developed independently in the Old World and the New and also independently in different geographical areas in both hemispheres.

Agricultural development occurred in association with a number of different crops, grains, tubers and tree crops but the one of the greatest relevance to us is rice, more specifically, paddy rice. The cultivation of rice uniquely among all crop plants lends itself to the culture of fish – a primitive form of aquaculture. The necessity of storing and using water provides an environment in which culture of a pond fish, such as the goldfish, becomes easy. Nutritionally any fish produced in the necessary reservoirs, ponds and irrigation ditches and even the rice paddies themselves would be a valuable addition to the diet. The high-protein fish and carbohydrate-rich rice would complement each other in producing an overall balance in the diet. The wild goldfish has long been a popular food fish in China and has been taken from the wild for this purpose for several millennia. At any given time, fish surplus to immediate requirements could have been kept for a longer or shorter period in accessible water bodies and taken as required. In permanent ponds and reservoirs resident populations could easily have become established. Goldfish are very adaptable and would have been well suited by such an environment and a basic system of aquaculture established with semi-domesticated populations of the goldfish. It may not have been necessary to supplement all of these populations by fish caught from the wild, in which case a certain level of inbreeding could have ensued. As a consequence of inbreeding recessive mutant genes, if they occur in the population, tend to become homozygous and come to be expressed. Any mutant genes present affecting colour could become apparent and fish showing colour variation would certainly attract attention and arouse interest. This may have aided their survival and if wild-type fish were preferentially taken for food this would confer a selective advantage on coloured variants which would tend to increase in the population. Where such populations thrived the attractive coloured fish might have been sold or bartered and thus been dispersed.

On the basis of what we know and what we can surmise a plausible hypothesis such as this can be produced. What evidence can we find to support it? Here we must turn to the writings of Hervey and Chen, especially the latter who has studied to great effect the earliest literature on the subject. Chen and Hervey consider literature going back to the first half of the first millennium and note reports of fish with red scales in the Qin dynasty period (AD 265–420) in a lake near Mount Lu and in the Red River near Shensi. Chen himself in 1931 saw in the Peking area coloured goldfish taken from the wild; it is possible that these were feral domesticates. It is also not absolutely certain that the 'red fish' were in fact goldfish; if they were not, however, then what were they? There is a strong possibility that they were. Even so, red or xanthic variants have been found in many Cyprinids over time and it is safe to conclude that the capacity to produce such variants is an inherent feature of fish not only in the Cyprinidae but in other families also.

Assuming that colour mutant genes were present in some wild populations, then if samples from such populations were taken and these succeeded in reproducing in captivity then it would only be a matter of time before red fish were produced as a result of inbreeding. Depending on the strength and efficiency of selection, sooner or later true breeding colonies of red-scaled fish could be established. This process could have been initiated as early as the T'ang dynasty (AD 618–906). The T'ang dynasty is considered by the Chinese and many sinologists to have been the pre-eminent Golden Age of Chinese culture and civilisation and it is quite likely that it was in this period that the preliminary phases of goldfish domestication took place.

It is at this point that we need to consider the religious dimension. Buddhism came to China from India in the first century of the first millennium and one of its more important tenets is respect for all forms of life. It was tacitly accepted by Buddhists that some sacrifice of animal life was inescapable if human life was to be sustained, but a symbolic gesture could be made by rescuing some potential food animals from their fate. In the case of food fish this took the form of establishing what Chen has called 'ponds of mercy', which we might be inclined to call fish sanctuaries. In the T'ang period there is a record of an order to construct 81 such ponds. The practice of releasing living creatures in this way began earlier but became an institution at this time and was known as 'fang sheng'. As Chen observed, 'The golden and yellow "chi" (i.e. goldfish) being rare in nature and fraught with mysterious significance would of course be among the first of the creatures to be set free.' There would of course be the advantage that their conspicuous coloration would make the fish highly visible and so their movements could be followed. It is well known that the movement of fish in calm waters is very soothing and must have been a considerable aid to contemplation and meditation, especially for Buddhist monks. Favoured sites for such ponds were in the grounds of temples and monasteries where the monks would act as guardians. Ideally all fish in such ponds would be fully protected.

The T'ang dynasty came to an end at the beginning of the tenth century, which was followed by an unsettled period before the Song (Sung) dynasty was established in the middle of the century in AD 960. Both Hervey and Chen cite numerous references in Chinese literature of this period to the goldfish in which it would appear that undoubted domestication occurred. At the end of the T'ang period there was probably a situation in which captive or semi-domesticated populations of goldfish were maintained with the beginnings perhaps of selection in favour of colour development. There does not appear to have been any sustained selection in favour of qualities as a food fish, as happened for example in Europe during the second millennium, for table quality in the common carp. It is interesting to note that the selection pressures on the goldfish were dictated by aesthetic considerations pure and simple, a remarkable situation for the time.

Domestication

In a very literal sense domestication took place in the period of the Song dynasty in that goldfish literally were brought into the home. After the fall of the T'ang dynasty there was a period in which the Chinese Empire fragmented (AD 907–960) before the Empire was once more consolidated by the first Song emperor. This dynasty lasted for just over 300 years before it too fell before the onslaught of Kublai Khan and the Mongols in 1279.

The major change which came about in Song times was the keeping of goldfish in private domestic ponds. The aquatic communities of the ponds of mercy were very mixed indeed and included fish species other than the goldfish, amphibia and aquatic reptiles such as terrapins. Chen believed that establishment of private ponds probably occurred during Southern Song times (when the Emperor moved south under pressure from Northern invaders), in Hangchow, where ponds of mercy had been established for a long time previously. The Emperor Chao Kou was apparently something of a goldfish enthusiast and in the year before he died, 1186, he ordered the collection of gold and silver fish to restock his ponds. Imperial favour could well have been a very potent influence in fostering and furthering the cult of the goldfish.

Keeping goldfish in single-species communities greatly facilitated the establishment of new variants, their maintenance and propagation. Their general welfare would also receive closer attention and the depredations of predators limited to a degree. Pond keepers would also become much more familiar with the reproductive processes of fish and learn how to rear progeny successfully. In any such venture as this some individuals through natural aptitude may have been more successful than others in rearing fish and found it profitable to engage in it as a hobby or even as a business operation. The ability to meet fashionable demands of society could well have become very profitable.

In the final years of the Song dynasty, society had become decadent and effete and no longer able to resist the pressure of the Mongol invaders from the North. The dynasty finally collapsed totally in AD 1279 and the Mongolian Yuan dynasty succeeded. Conquering an empire is one thing, ruling it over the long term is another; the Yuan dynasty was relatively short-lived and in turn it was succeeded by the Ming dynasty in 1368. The Yuan dynasty had lasted a mere 89 years, AD 1279–1368.

Of all Chinese Imperial dynasties the Ming is arguably the best known in the West. This is due largely to the fame of the ceramics produced in the period, which are now the most highly prized and valued of any. The question of this era being the Golden Age of Chinese ceramics had a most important impact on the cult of the goldfish. The production of ceramics spanning the whole range of quality from the functional to the superb and in a size range from the small to the enormous meant that it was possible to produce vessels large enough to accommodate goldfish indefinitely in which they could not only live but breed. The keeping of goldfish in Song times in private ponds gave a measure of control over predation and breeding, which was taken to the ultimate during the Ming dynasty when total control over predation and breeding became practicable. Rearing fry in a protected environment eliminated much of the natural selection which would have operated in an open environment such as a pond or river. As a result, mutant forms came to light which, while uncompetitive in a natural or artificial pond environment, were quite viable in the protected environment provided by a large bowl or ceramic tub. Control of breeding also became easier, parents could be selected and removed after spawning and the fertilised eggs hatched in an optimal environment. Rearing of fry could be supervised closely and appropriate use made of Daphnia and other nutritious live foods in the feeding and rearing of goldfish.

Chen makes the interesting comment that from the time of the collapse of the Song dynasty in 1276 until 1546 there is little literary comment on goldfish matters. Since the Ming dynasty itself came to an end in 1644, the important developments in goldfish culture were concentrated in the latter half of the period. Presumably culture in ponds continued until the watershed year of 1546 when the use of earthenware vessels came to the fore. Initially these bowls were probably stocked with a representative sample of pond life including fish. Housing in such vessels may initially only been temporary, then becoming permanent with the residents not only surviving but also breeding. By careful selection of parents, interesting novel types could be fixed and propagated. With the inbreeding that fixation of novel types implies, further novelties might arise at the same time as desirable phenotypes were exposed to selection designed to enhance their phenotypes. The era of aquarium culture of the goldfish can be dated from 1546 when, judging by the amount of literature, including much poetry on the goldfish, there was a tremendous upsurge of interest.

We need now to consider what was the upshot of this final century of intense interest in the goldfish in Ming times. An important point to note is that when goldfish culture was centred on the pond, keeping them was the prerogative of the wealthy or of institutions such as temples or monasteries. The situation was probably not dissimilar to that of Koi culture at the present time, when to make a proper job of it requires the expenditure of a great deal of money and time and also takes up a great deal of space. When it was found that goldfish could be kept more or less indefinitely in good health in relatively small containers then the goldfish was poised to become the pet of the masses, which subsequently came to pass not only in China but elsewhere in the world.

As far as new variants are concerned there were two different shades of red recognised, the 'fire-fish' and the 'cinnabar fish' which had been established in the era of pond culture; subsequently a profusion of colour variants were recorded, many of which were dappled. These were often given poetic names such as 'golden helmet', 'golden saddle', 'stork's pearl' and 'brocaded back', these are very similar to names which have been given by the Japanese to Koi varieties. What is of more interest are reports of numerous 'tails', three-tail, four-tail, five-tail, seven-tail and nine-tail variants, as well as bulging eyes and the short body. In addition there are reports of rather rare transparent fish in which 'when you look at it closely you can see both the stomach and intestine'. A range of variants of such transparent fish were described which Chen sensibly points out can be described as 'transparent and mottled' or, as we might now say, 'calico'.

The group of 'three, four, five, seven and nine tails' is of interest and embraces what we would now call 'double-tails' or 'twin-tails'. In these descriptions lobes of fins are considered as 'tails'; it is possible that five, seven and nine lobes might arise as a result of split fins, inclusion of anals and ventral fins, or occasionally production of aberrant lobes. The mutation responsible for the duplication of the caudal (and anal) fins is arguably the most seminal change in the production of what we now call 'fancy' goldfish. Taking this in conjunction with the development of protuberant eyes and short bodies, the scene was clearly set for the explosive development of goldfish varieties which followed. The critical date for the first record of the double tail was 1579; Chen considered that it probably arose between 1521 and 1579. Chen also concluded that some of the fins described were long.

Diversification

Chen has called the period from the beginning of the Qing (Ch'ing) dynasty (1644–1912) to the time he wrote his 1954 paper *The Era of Conscious Artificial Selection*. This is when the variation which had been generated largely in the latter half of the Ming period became patterned into the exotic forms we can recognise today. The results of this became apparent especially in the eighteenth and nineteenth centuries. By 1849 it was already clear to Chen from the sources he quoted that considerable care was taken in the selection of brood stock and that culling rates of 90% operated, so that only the very best were retained for further use. Chen also mentioned that towards the end of the nineteenth century two new types are recorded, the 'inky dragon-eye' (? Moor) and the 'lion-head' (1893). A decade or so later 'the sky-gazing dragon' (celestial), the 'tiger-head' (goosehead) and the 'narial bouquet' were recorded in 1904 (see Fig. 2.1). Chen finally noted that by 1925 the pearlscale, bubble-eye, out-turned operculum, the blue fish and the purple fish (i.e. brown or chocolate) varieties were established.

Hervey and Hems provide evidence which is in the main consistent with Chen's propositions, frequently from different literary sources. Among the significant dates recorded are that by 1590 'we read of fish with three tails or four tails (the Chinese counted each lobe as a

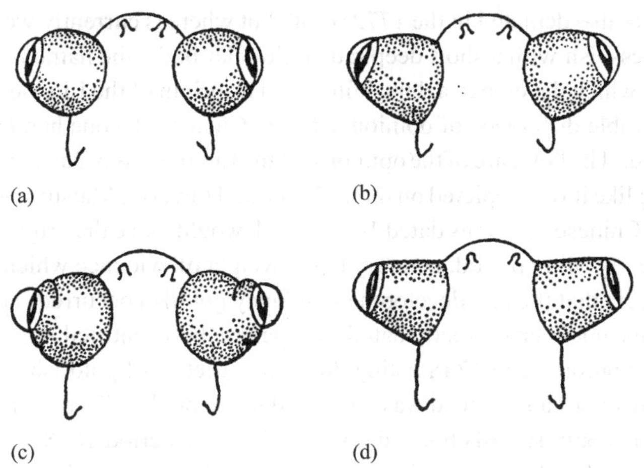

Fig. 2.1 Forms of telescope eye: A, spherical; B, ovoid; C, segmented sphere; D, truncated cone (Hervey & Hems 1948).

tail)' which were recorded as being a recent development by fanciers and that they were not in existence in the early part of the century. The origin of short-bodied fish was noted at the same time, and an association between short bodies and double tails. These observations were recorded in the *K'ao P'ou Yu Shih* (1590) which includes an uncertain or ambiguous reference to telescope eyes. Telescope eyes were certainly known in the mid-eighteenth century when numerous examples are depicted in the de Sauvigny scroll of 1772. At about the same time (1765) Baster (a Dutchman) noted fish with bulging eyes 'unusually so in some specimens'. The telescope fish became quite well known in Europe in the latter part of the nineteenth century (Carbonnier 1872).

We have already noted the diversification of colour which had been established during the Song and Ming periods. The occurrence of transparent scaled goldfish was recorded in 1688 by Ch'ên Hao-tzû 'who tells us of the many-coloured variegated fish bred in the Phoenix Well' in Kiangsi and in two locations in the Hangchow area and one in Chekiang. It is curious that this character did not become more widespread in the eighteenth and nineteenth centuries. It persisted in the Calico Telescope, from which it was transferred by crossing most notably to the Shubunkin and to most twin-tailed varieties in the course of the twentieth century. The origin of the dorsal-less character is of considerable interest (especially to Ranchu enthusiasts!). Hervey and Hems reproduce an illustration dated 1726 showing three twin-tailed fish, one of which has no dorsal fin. Such fish were known in Europe in 1760. Edwards refers to fish 'with little risings on the middle of the back in the place where the fins generally are'; any breeder of dorsal-less varieties knows exactly what is referred to!

The origin of the Celestial appears to have been in the eighteenth century; there are one or two dorsal-less telescope fish depicted in the 1772 scroll whose eyes seem to have a decidedly upward cast. The authors mention that it might have had a Korean origin. It is perhaps safer to conclude that it was of Chinese origin though favoured by Koreans. The possibility of a Korean origin cannot be totally discounted; if it were to be substantiated it would have been quite a unique contribution to goldfish culture.

The Eggfish is also depicted in the 1772 scroll, but whereas currently we use the name to refer to dorsal-less fish with a short deep rather globular body, the name is used at the time to include those which developed a dorsal fin also. The origin of the Lionhead is perhaps the only area of possible difference of opinion between Chen on the one hand and Hervey and Hems on the other. The latter are of the opinion that the Lionhead is of Japanese origin, largely because nothing like it was depicted on the 1772 scroll. However, Matsui (1971) noted that it was depicted in Chinese paintings dated 1429, which would place the origin of the Lionhead as being in the early Ming period, contrary to the weight of evidence which indicates that it was later in this era that the real development of fancy goldfish occurred. This suggests that the paintings have not been correctly dated. A Japanese book entitled *A Notebook on Raising Goldfish* was produced in 1748 listing different varieties of goldfish while, in the year following, a drawing of a Lionhead was published in the book *A Record of the Mysteries of the Goldfish*. Chen only records hooded goldfish from the period 1848–1925 in which the Lionhead, Tigerhead and Goosehead were produced. These uncertainties notwithstanding there can be little doubt of the enormous contribution which Japanese breeders have madc in the development of the Lionhead.

Perhaps even more controversial is the origin of the Oranda. A commonly held view is that it arose from a cross between the Ryukin and Lionhead. Hervey and Hems stated that the Oranda was first bred at Koriyama or Osaka in 1840 from a Ryukin (Veiltail) × Lionhead cross. Matsui disputed this claim and recorded the publication of a drawing of a Dutch Lionhead (Oranda Shishigashira) by Hirokawa in Nagasaki in 1800. At a show in 1883 an Oranda was exhibited with an explanation that this type had been bred for about 30 years in Japan but that its origin was uncertain. Matsui also pointed out that whereas in the Lionhead there can be variation in development of the dorsal fin, from complete absence to production of vestigial and rudimentary fins, no such segregation occurs in the Oranda. It is possible that Oranda-like fish were produced in two ways, one by the hybridisation considered, the other by the establishment of a line of hooded Fringetail. It is a matter of fairly common observation that with increasing age, fish of the Fringetail family can show hypertrophic skin growth on the cranium. This can even be seen in the magnificent coloured frontispiece of Innes' *Goldfish Varieties*. Equally it is possible that the absence of the dorsal fin could have evolved independently in different goldfish lineages. If one raises large progenies without severe culling, fish with incomplete development of the dorsal fin are encountered not infrequently. Selection of such individuals for breeding and subsequent further selection could well ultimately result in new dorsal-less strains. A comparison between Lionheads and Bubble-eyes shows that the most significant features they have in common are the lack of dorsal fin and possession of twin-tails. Body conformation and the nature of the finnage are quite different.

At about the turn of the last century, the focus of activity in goldfish variety development shifted to a substantial degree from China to Japan. Basically Chinese stocks were taken from about 1500 and through selection modified (or 'improved'). Certainly there were features which were favoured by selection among Japanese breeders. If one examines the fish depicted on the de Sauvigny scroll of 1772, all are short-finned. Long fins have been favoured in many Japanese varieties, most notably the Fringetail, and the production of varieties within this family such as thc 'Ribbontail' has occurred; the name speaks for itself. Perhaps the distinctive feature of Japanese breeding activity was the use of deliberate hybridisation to create new varieties; the greatest success was in the production of the Shubunkin, one might almost say the fish which won the West! Matsui documented its origins from a complex cross carried out by Kichigoro Akiyama I. A Calico Telescope was crossed with a 'Japanese Golden' that had a Carp tail and further with a Scarlet Crucian (Hibuna). The name Shubunkin (Red Marked Calico) was given by Matsubara who claimed to have reproduced the same result from the Calico Telescope × Japanese Golden cross. Matsui argued that this would have produced a Calico Wakin, not a Shubunkin. He carried out the cross Calico Telescope × Crucian (wild goldfish). He found that about half of the progeny were of the normal shape, the proportion of transparent scales was about a quarter, while all the progeny had single tails and normal eyes. The original cross involving three lines was a roundabout way of reaching the objectives which Matsui achieved more effectively. Coloration was very varied; a proportion were yellow and black while a substantial number had more complicated patterns. Presumably the desirable red and blue combination required further selection and crosses between selected parents. The length of the fins exceeded that of the standard or wild parent. The form of the Shubunkin has remained basically similar in Japan and the East for the whole century but has been 'developed' further in Great Britain and the United States, about which more later.

An intriguing fish mentioned and illustrated in Matsui's books is the Yamagata goldfish which is essentially a twin-tailed Comet. Apparently it was developed as a hardy variety for the Yamagata area which has a severe winter climate. It is not widely known outside Japan but could easily be produced from a cross between a Wakin and a Comet. It could be an interesting pond fish.

Most hybridisations carried out in Japan and elsewhere have involved one of the parents of the Shubunkin – the Calico Telescope – and most of the original metallic scaled varieties; these have been very successful, the Calico Oranda can, for example, be a superb fish. Some others have been notably less successful, the Kinranshi or Golden Bleary-eye (a most inauspicious name!) was produced from a Wakin × Ranchu cross made by Kichigoro Akiyama in 1902, the breeder who produced the Shubunkin. It apparently had the ability to produce attractive golden reflections but never achieved any great popularity although it is noted in some quite recent works (Appendix III, Watanabe 1988).

Finally, brief mention should be made of the Watonai, the result of the cross between the Fringetail and the Wakin. The result is a fish which the Western enthusiast would consider to be a Fantail. It could be regarded as a reconstruction of a stage in the evolution of the Ryukin or Fringetail from the Wakin.

The next significant developments in goldfish culture came at around the turn of the century in the United States. The earliest record of the introduction of the goldfish to the USA was in 1874 and credited to Admiral Ammon. It is hard to believe that this was actually the first introduction of this fish to North America. In any event the first significant involvement of American goldfish fanciers with development of the goldfish varieties was in the 1880s (a decade or so after the first recorded introduction of the goldfish to the United States). Hugo Mullertt propagated the Comet goldfish from selections taken from the ponds of the Fish Commission in Washington. This was followed around the turn of the century by the Veiltail which has an interesting history. The story began in 1893 at the Chicago World's Fair. A group of Fringetail goldfish were intended exhibits but fell sick. The survivors were rescued by William P. Seal and subsequently came into the hands of a Mr Barrett who was the initiator of the

Philadelphia Veiltail strain. This strain was founded in the main on one individual, the famous 'World's Fair Fish' which survived for 15 years.

The most interesting feature of this strain is that its characteristics were initially surprisingly definitive; almost the only significant departure from the original standard is the modern preference for a broadtail rather than the somewhat bifurcated tail of the original. The latter is a characteristic feature of the original Fringetail stock. Philadelphia Veiltails were eagerly taken up by European fanciers, mostly British, with whom it has remained a firm favourite. The Veiltail in Europe which was kept in both Germany and Russia was virtually wiped out in the course of World War II but survived in Britain.

The Shubunkin became very popular in America and was subjected to a selection regime designed to increase the length of finnage. This produced a form which is sometimes called the American Shubunkin or Calico Comet. Work is continuing in the United States in an attempt to produce a type of Shubunkin which breeds true. This is an interesting development which will be discussed further in a later chapter. Further development of the Shubunkin took place between the wars in Britain where the first introductions occurred in the early 1920s. From these L. B. Katterns and A. Derham are credited with developing the London Shubunkin. This has the body form and fin characteristics of the common goldfish and has often been considered as merely a Calico Common Goldfish. It is possible that, in the same way that Calico Veiltails have been produced by crossing and backcrossing, the London Shubunkin might have been produced in a parallel fashion. The London Shubunkin still has its devotees and some outstanding examples have been produced in recent years by Messrs W. Leach and A. Ratcliffe and Mrs P. Whittington. However, in popular esteem it must yield the palm to the Bristol Shubunkin which is, by any standards, a remarkable fish. Its initial development between the wars can be credited to members of the Bristol Aquarists Society, which was devoted to the keeping and breeding of coldwater fish. By the beginning of World War II the Bristol Shubunkin was regarded and recognised as a distinctive goldfish variety.

The development and evolution of new variants of goldfish is still continuing, mainly in China. Many of these in essence are ringing changes on established themes, but it is hoped that through a better understanding and application of genetic principles we can more easily conserve and maintain the diversity of goldfish forms and also augment it.

The political dimension

The logical point in history to start the consideration of the implications of politics on domestication and development of the goldfish is the time when China was first unified during the short-lived Qin or Ch'in dynasty (221–209 BC). After a very brief interregnum this was followed by the Han dynasty which lasted for over 400 years (206 BC– AD 220). In this period Buddhism came to China. This was followed by a period in which China was ruled as Three Kingdoms (AD 220–598). Then followed the short-lived Sui dynasty (AD 596–618) and the T'ang dynasty (AD 618–906) which saw remarkable cultural developments in arts, crafts and technology and regarded by many Chinese historians as the major peak of Chinese civilisation, truly a Golden Age. It was at this time, as we have seen, that the scene was set for goldfish domestication and development initially by the proliferation of ponds of mercy in proximity to Buddhist temples and monasteries. The presence of goldfish in these ponds was a direct consequence of their common use as a food fish. The coloured fish whose occurrence was reported sporadically as early as the Tsin dynasty (AD 265–419, one of the Three Kingdoms) would have been favoured for rescue, being considered as auspicious in themselves and readily visible in the ponds of mercy as an aid to reflection and meditation for the monks.

The end of the T'ang dynasty was followed by a period in which the rule of the country was fragmented, which lasted from AD 907–960 until unification was achieved under the Song (Sung) dynasty (AD 960–1279) when there is general agreement that full domestication of the goldfish occurred. Unlike the common carp in which distinctive domesticated forms were developed for food purposes, the mirror and leather carp most notably, the domesticated forms of goldfish were developed exclusively for aesthetic and ornamental purposes. Goldfish used as food were essentially of the wild type; there was also a reported reluctance to use coloured fish for this purpose. During the Song period China was attacked from the North and the Chinese court fled south and established a capital in Hangchow. The Song dynastic period comprises the Northern Song period 960–1127 and the Southern Song period 1128–1279. The latter period was that in which the most rapid early development of the goldfish apparently took place. Final collapse of the Song dynasty took place with the onset of the Mongolian invasions of Kublai Khan. The Mongols reunified the country by overcoming both the Chin invaders who brought about the retreat south of the Song culture and the Song rulers too. They established the Yuan dynasty which ruled for approximately 100 years. It was the first foreign dynasty to rule China.

The ultimate collapse of the Yuan dynasty and its succession by Ming rule brought about another Golden Age of Chinese culture, celebrated in the West for the very highly prized ceramics of the period. It was at this time that the foundations were laid and the first steps taken leading to development of the goldfish as we know it today. The downfall of the Ming dynasty was brought about by invaders from Manchuria who established the Qing, Ch'ing or Manchu dynasty which was in power from 1644 until 1911 with the establishment of the Republic formalised in 1912. The Republic of China itself fell in 1949 in the Communist Revolution and the People's Republic of China was established in that year. Several very significant events (or series of events) took place in the nineteenth and twentieth centuries. Western mercantile policies which had been initiated as far back as the sixteenth century produced a crescendo of highly aggressive and intrusive efforts to 'open up the country to trade'. From the time of the Opium War in 1839, the attempts of the Chinese to resist these efforts and the continuing pressures of western commercial interests resulted in the Taiping Rebellion of 1851–64 and the Boxer Rebellion of 1899–1900. In China the effects on goldfish development were surprisingly small, in fact, according to Chen, in the period 1848–1925 ten new varieties were developed.

For China, politically speaking, the twentieth century has been little improvement on the nineteenth as far as general peace and harmony are concerned. The establishment of the Republic was followed by the establishment of the Guomindang or Kuomintang and the Communist Party factions whose struggles continued until 1931 when the Second Sino-Japanese War started which lasted until 1945. The end of World War II was followed by a civil war lasting until 1949 with the fall of the Guomindang. In such troubled times it was unlikely that goldfish culture would burgeon, but it survived. Unfortunately worse was to follow. The decade beginning in 1966 was the period of the Cultural Revolution in which traditional culture which had hitherto been respected by the Communist Party came under attack by the Red Guards.

This movement, which can be likened to a cultural Khmer Rouge, brought about untold destruction to Chinese art and culture before it came to an end in the mid-1970s. The harm done to goldfish stocks is incalculable. The cult of the goldfish in the eyes of the Red Guard movement was politically incorrect and among many other harmless and even laudable activities were to be ruthlessly suppressed. This was arguably the worst catastrophe ever to befall goldfish culture, largely because it came about over the whole country simultaneously. However, it is a great tribute to the Chinese people and the resilience of their culture in general and that of the goldfish in particular that it survived, and we can reasonably hope for better times ahead. Over the past two decades the Chinese goldfish breeders have shown magnificently that they have not lost the capacity to surprise and delight goldfish fanciers the world over. The future hopes of the goldfish fancier still rest as ever in China.

In his general discussion of goldfish domestication, Chen noted that the appearance of new goldfish varieties did not occur in a steady pattern from his critical date of 1163. In the years between 1163 and 1925, a total of 762 years, there were three periods which were particularly productive of new variations. The first of these was 113 years from 1163 to the end of the Southern Song period (1128–1279). The second was 97 years in the late Ming period, between 1547 and 1644. The final period of 1848 until 1925 covered the concluding years of the Qing dynasty and the early years of the Republic of China, in fact until almost the eve of the troubles that were to beset China for much of the twentieth century at the hands of the Japanese and internal political factions. From this Chen concludes that 'the demand for novel varieties of goldfish in society calls for men who breed, select, and record the fish. When such a demand in society is absent, the discovery of new methods of culture and the selection of new varieties are suspended.' Chen's first two points, that it is necessary to have men who breed and select fish, cannot be denied. Recording is quite another matter. Those who record are not necessarily those who breed. The goldfish survived the apparently unproductive periods and as far as we can make out did not lose ground in the process. My own experience in attempting to record the activities of breeders suggests that, in at least some and perhaps even in the majority of cases, this may well be the last thing they want to happen. Anecdotal evidence suggests that the great French water-lily breeder Marliac would not even tell his own sons all the secrets of his success!

There is a significant pattern in the timing of these very productive periods which were in the final years of the Song, Ming and Qing dynasties. Chen pointed out that in the early years of dynasties, the population would be settling down and adjusting to the political change and it would take time for a high level of cultural activity to be re-established. I believe that the people most affected by dynastic changes would be emperors, courtiers and officials. Farmers, artisans and craftsmen, though by no means unaffected, would have been needed by the new regime and would be in the best position to maintain the stocks of goldfish which they had developed and prized. It is not uncommon in the world of animal breeding to find that rather few of the breeders of outstanding animals are suave, educated gentry with a good scientific background, except where the animals concerned have been farm livestock and horses. While Bakewell who established the Longhorn cattle breed is well known, who knows the initiators of many of the breeds of the smaller domestic pet mammals and birds? Darwin is the best known pigeon breeder but how many others are known to fame? It is reasonable to think that the variety of goldfish was maintained as a folk culture activity and that periodically the well-to-do and leisured classes succumbed to fashionable crazes among which goldfish figured. We have only to consider the present crazes for Bonsai and Koi to appreciate what can happen. This situation creates the search for novelties, the producers of these at the present times may be known through books and magazines but who are often rather self-effacing and not publicity seekers. During the twentieth century when scientists became interested in the evolution and development of goldfish varieties, breeding experiments were described and parentage of crosses published. When new varieties are produced commercially the modes of origin of these are, for understandable reasons, not publicised and this equally understandably gives rise to conjecture and speculation as to how they were produced.

We are now in a position where we can attempt to answer our initial questions, when was the goldfish domesticated, where, by whom, how and why this came about. The goldfish was domesticated in China towards the end of the first millennium. The scene might actually have been set for this somewhat earlier in the Tang dynasty but it had certainly come about in the Song period. It seems probable that the initial domestication was due to the efforts of farmers who caught the goldfish initially for food but were attracted by the occasional coloured variants which arose from time to time and which were retained and not consumed. This was assisted by Buddhist monks who provided refuges for animals including fish spared from slaughter, which in all probability included coloured goldfish. The reason why domestication is followed by the establishment of the very diverse forms we see today probably stemmed from the innate tendency for human creativity to find expression. The production of finely worked beautiful flint tools provides an early illustration of human creativity. The amount of effort put into some of these artefacts was well above and beyond what was needed to produce a serviceable tool. Aspiration towards perfection is what characterises people of an artistic bent and has also found expression in the work of those who have produced the astonishing array of beautiful, interesting and useful variants which we can readily see among domesticated animals and plants, which make civilised life possible.

Artefacts as evidence in tracing development of goldfish varieties

The value of artefacts in tracing development of variation in domestication depends in the first place on the accuracy with which their form is reproduced and secondly the accuracy with which they can be dated. In the case of Chinese and Japanese works of art, the form of goldfish has often been produced with astonishing accuracy. The features of greatest concern to the student of goldfish evolution have been reproduced in depictions in the decoration of ceramics, for example, from the Ming period. However, the most remarkable work of art devoted to the goldfish was the famous scroll of 1772 depicting 92 goldfish. This was produced by Billardon de Sauvigny as part of his *Histoire Naturelle des Dorades de la Chine* published in Paris in 1780. The scroll was painted by Chinese artists and a wide range of goldfish forms are represented. George Hervey in 1950 published a commentary and translation of de Sauvigny's manuscript which gives an unusually full perspective of the range of variability apparent at the time. It is quite unique in this respect and an absolutely unrivalled source of information. One might well conclude that it provides some insight into the kind of variants goldfish breeders had at their disposal in the early days of Chen's *Era of Conscious Artificial Selection*.

It is readily apparent that two distinct types of body, elongated (Fig. 2.2) and short (Fig. 2.3), can be recognised, duplication of the caudal fin is complete in some individuals and

Fig. 2.2 Long-bodied Telescope (eighteenth-century Chinese scroll). Note completely duplicated caudals.

Fig. 2.3 Short-bodied Telescope (eighteenth-century Chinese scroll).

the telescope eye is in evidence. What is not apparent is any individual fish with long fins. This highlights a major difficulty in reconstructing evolutionary sequences, while it is possible to establish with a degree of certainty the presence of a character, uncertainty attaches itself to any presumption of absence of any specific characters (Figs 2.2, 2.3).

Dissemination of the goldfish

Documentation of the dissemination of the goldfish in the Far East is not very extensive. It seems probable that it was taken to Indo-China, Korea and other neighbouring countries before the time of its introduction to Japan at the very beginning of the sixteenth century. It is thought that not until the eighteenth century did serious goldfish breeding begin in Japan (Hervey & Hems 1948). It is possible that goldfish came to Japan by way of Korea, if the Lionhead truly deserves the name in Japanese of Korean goldfish, and via the Ryukyu Islands if the name Ryukin has any real significance. Matsui (1934) suggests that introductions to Japan from China occurred in 1502 or 1620. It is highly probable that after initial introductions in the sixteenth or seventeenth centuries (or both) that periods of introductions occurred subsequently. Matsui identified four basic prototypes of Chinese origin: the Hibuna (red Crucian carp) Wakin, Ranchu, Ryukin and Demekin. He suggested that Japanese varieties arose from these prototypes by selection and hybridisation.

The first introduction of the goldfish to Europe was to Portugal. The Portuguese had established a port in Macao in 1557 and it seems likely that the first live goldfish to be sent to Europe came to Portugal, perhaps in the seventeenth century. There are records of goldfish in England as early as Pepys' diary (1665) in which he refers to foreign, finely marked fish being kept indefinitely in glasses of water. The source of the English goldfish was Macao in all probability. Certainly by the middle of the century they were well established and recorded as occurring in ponds at Vauxhall. Horace Walpole (1717–1797) seems to have been the archetypical goldfish enthusiast who established a thriving colony on his estate and was reported as constantly giving them to his friends. During the course of the eighteenth century the goldfish had been distributed generally over Europe and by 1791 was clearly established in Russia. In was known to Linnaeus by 1740 and there are records of it during the century also from France, Holland and Germany. It is also interesting to note that Thomas Grey wrote his 'Ode to a favourite cat drowned in a tub of goldfishes' in the closing years of the eighteenth century.

The development of novel forms of goldfish outside China did not apparently get under way for a considerable time after the initial introduction to Japan. It is thought that in the late seventeenth century goldfish keeping was prescribed as a degenerate pastime and only in the eighteenth century did it begin to come into its own. In the United States by contrast, if one accepts the 1874 date of introduction as reliable, within only two or three decades the Veiltail was developed followed by the Comet. In Britain three centuries elapsed before the changes were rung on the Shubunkin theme.

Chapter 3

Goldfish Varieties – a Review of Literature

During the course of the second millennium the goldfish has produced a very wide range of variants, many of which survive to the present day. It is quite apparent from reading literature on the subject from the past century that our present varieties do not necessarily represent the full range of variant forms which have arisen. Our present perception of goldfish varieties goes back essentially to the work of William T. Innes and his classic work *Goldfish Varieties and Tropical Aquarium Fishes* which was first published in 1917 and ran through many editions until 1932. In 1935 an expanded version of the section on tropical fish was published as a new work *Exotic Aquarium Fishes* which has been revised and reprinted many times since and has become the classic volume on tropical fish keeping. After the end of World War II the goldfish section was revised and published as *Goldfish Varieties and Water Gardens*.

Innes' work did not come out of a vacuum; in the previous decade a work entitled *Goldfish Breeds and Other Aquarium Fish*es was published (1908). This is a most interesting book and can be considered as the initial trailblazer. This work by H. T. Wolf was one of the earliest if not the earliest to attempt a comprehensive description of contemporary goldfish varieties. It is interesting to compare the listings given in the more important books during the course of the twentieth century; it gives some reflection on what has been popular and available at various times. Intriguingly over this relatively short period some varieties have made a brief appearance then disappeared, some for a time and some apparently permanently. The names also undergo change; varieties can disappear from view under one name and reappear under another. Goldfish authors naturally write on the basis of their own experience and some may well encounter variants which others never see.

While reference will be made largely to goldfish literature in English, there are references to goldfish varieties in French and German literature, for example, Pouchet (1870). However, as these accounts are sometimes not illustrated it may not be easy to visualise the forms described. It seems sensible therefore to confine the discussion largely to those texts in the English language in which every form described is also illustrated. In passing, it should be noted that there are illustrated references to goldfish in nineteenth-century literature. A good example is to be found in the Rev. W. Houghton's *British Freshwater Fishes* (p.57) published in 1879 which depicts two colour variants, one predominantly red with a silver belly, the other with a red back and silver flanks and belly. He also indicates briefly the range of fancy goldfish currently available, most notably those in which the dorsal fin may be reduced or absent and in which the anal fin and caudal fin may be duplicated. He also refers to what appears to him to be the strangest of all, the very large and protruding eyes of the telescope fish. The only other variety he mentions by name is the 'Japanese fan-tail'.

At about this time there were a number of books published on aquarium keeping in which goldfish are mentioned, namely by J. E. Taylor (1881) and the Rev. G. C. Bateman (1890). The latter devotes a chapter to goldfish varieties and their breeding. His comment on the Japanese Fan-tail is interesting; he noted that the tail was frequently double, having the appearance of an inverted Y, what we would call a 'tripod tail'. When the tail is single 'the fish is sometimes called the Comet Fish'. In his notes on the Telescope Fish he recorded that the eyes were forward facing. At the present time this is seen as a phase in the development of the Celestial. There is a drawing of this fish which has a fully duplicated tail, a moderately deep body and very definitely forward facing eyes, and the dorsal fin is present. He reserved his greatest admiration for the Japanese Fringetail. He commented that perfect specimens had been sold for 'five guineas and even twenty times their weight in gold'. The colour range was wide, the commonest being red body and white fins but sometimes white body with red fins. It is difficult to be absolutely sure to what presentday varieties his 'Mottled Beauty', 'Blue' and 'Superb' correspond, as he only describes colour. The first could be a calico fish, the second a blue metallic while the last, a large black and scarlet fish, may have been a fish in the course of the colour change. However, as has been noted elsewhere, the colour change in mature fish may be very slow and prolonged, perhaps never going to completion.

It is not surprising that the goldfish has engaged the attention of eminent scientists. Charles Darwin in his work *The Variation of Animals and Plants under Domestication* (1868) devoted a page and a half to the goldfish. He was unsure whether to consider varieties such as 'triple tail-fins, etc.' as such or as monstrosities. He did conclude 'that it is difficult to draw a distinct line between a variation and a monstrosity'. The difficulty is perhaps still with us. He recorded variants lacking a dorsal fin, with double anals and triple tails. Darwin voiced the opinion that duplication of the caudal fin might possibly be at the expense of fin development elsewhere but cited a report from Madrid of a fish with both a dorsal fin and triple tail. A variety with a humped-back (? Ryukin) is recorded and a remarkable fish with 'the fleshy part of the tail as if entirely cut away; the caudal fin being set on a little behind the dorsal and immediately above the anal'. Both the caudal and anal fins were double, the body was globular in shape and the eyes enormously large and protuberant.

William Bateson in his *Materials for the Study of Variation* (1894) commented on fish obtained from Japan. Three distinct breeds he believed were maintained there, the Wakin, the Maruko or Ranchiu and the Riukin. These are names with which we are familiar today, allowing for some slight variation in spellings. He considered the 'Riukin' to have the most beautiful tail which was very large and often longer than the body. He stated that 'Goldfish breeders of the present day can freely produce the "Riukin" or "Maruko" from the "Wakin". Various intermediate forms between the above-mentioned breeds exist.' While today we would accept that goldfish generally do not breed true, we would find difficulty in accepting the idea that a Ryukin could be selected easily from a Wakin progeny. Bateson gives a very good and concise illustrated description of single, double and intermediate forms of the caudal fin. He also quoted a report that twin-tailed fish had been found in running streams. It is also important to note that the goldfish described by Linnaeus was a twin-tailed form and that this type was well known in Europe in the eighteenth century.

H. T. Wolf (1908) *Goldfish Breeds*

It was in the early twentieth century that specialised treatments of the goldfish began to appear, among which was Herman T. Wolf's pioneer work entitled *Goldfish Breeds*, published in Philadelphia in 1908. It covered not only the care and husbandry of goldfish but also selected North American freshwater fish, invertebrates and plants. He also included chapters on marine aquaria and terraria. For our purposes we need only review his coverage of goldfish varieties (or breeds, as he termed them). These are listed in Table 3.1 with a commentary on his treatments of individual varieties; this pattern will be repeated for those of other authors and these views will be subsumed in subsequent sections devoted to the major contemporary varieties.

1 The Common Goldfish (Figs 3.1a, 3.1b)

Wolf recognises two forms of Common Goldfish which he calls the American and the European. The latter is more slender and elongated in body form than the former. The basis of this distinction is not clear; subsequent descriptions closely approximate to the American rather than the European form. It is possible that the description of the latter was based on a small and atypical sample. From the published drawing, the European variety appears to be quite an elegant variant. Colour variants noted are white, silvery-grey, olivate, golden or orange yellow, red and brown, singly or in combination. Some whites are reportedly albinos with red eyes.

Size recorded is up to 16 inches in length and longevity 12–16 years commonly; exceptionally, however, both length and longevity may exceed these figures.

2 Japanese Comet Goldfish (Figs 3.2a, 3.2b)

The Comet, Wolf claims, was produced by crossing the Japanese Fringetail and Common Goldfish and first reported in the United States in 1872. The body form is relatively slender and the fins are elongated. In his drawing the height of the dorsal exceeds the body depth while the length of pectoral, pelvic and anal fins is approximately equal to depth of body. The dorsal and ventral margins of the caudal fin are approximately equal in length to the body. Tail fins noted were either strongly bifurcated or full broad tails. Scale-less (i.e. transparently scaled) Comets are also mentioned. The example depicted has a tail-fin length (along dorsal or ventral

Table 3.1 Wolf's Goldfish Variety Listing (1908).

1 Common Goldfish (Figs 3.1a, b)	11 Chinese Moor Telescope
2 Japanese Comet (Figs 3.2a, b)	12 Chinese Piebald or Tiger Telescope (Figs 3.6a, b)
3 Japanese Fringetail (Fig. 3.3)	13 Chinese Lettered Telescope
4 Japanese Fantail	14 Chinese Blue Telescope
5 Japanese Nymph	15 Chinese Celestial Telescope
6 Japanese Hooded or Lion-headed Goldfish (Fig. 3.4)	16 Chinese Eggfish (Fig. 3.7)
7 The Japanese Barnacled or Paradise Goldfish	17 Chinese Tumbler (Fig. 3.8)
8 Chinese Telescope Fish (Fig. 3.5)	18 Agard's Wonder (Fig. 3.9)
9 Chinese Mottled or Variegated Telescope	19 Lawson's White Rat (Figs 3.10a, b)
10 Chinese Fringetail Telescope	

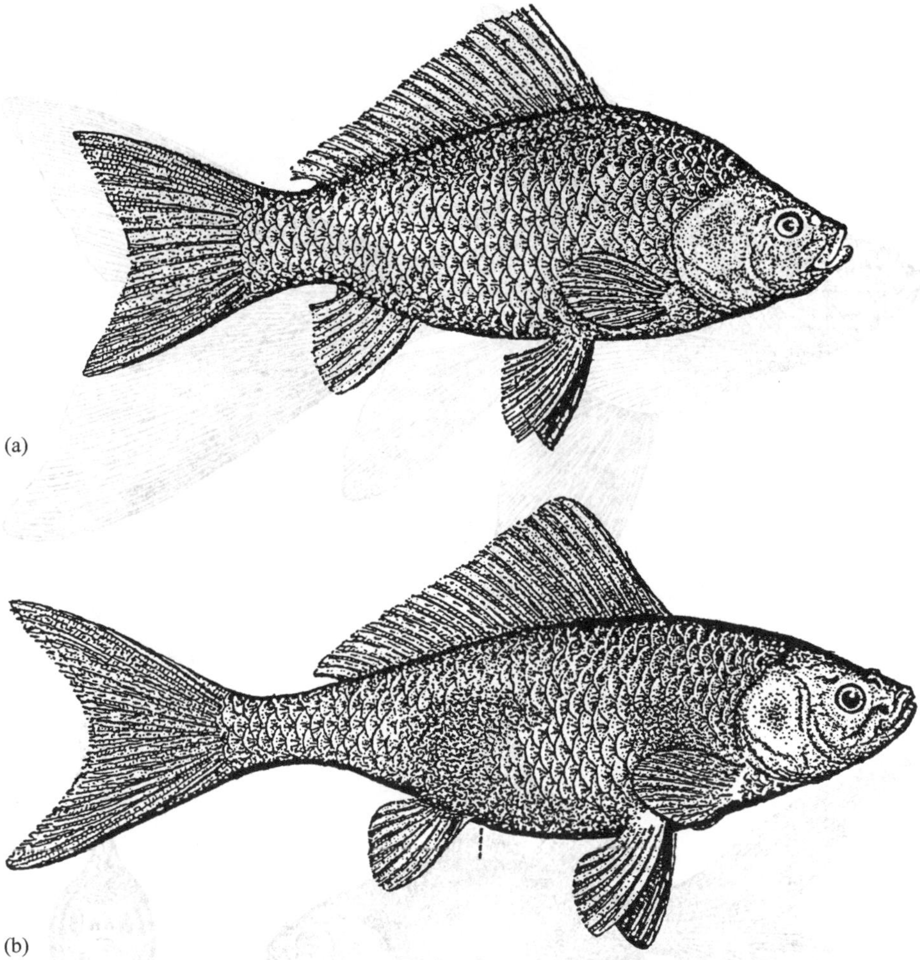

(a)

(b)

Fig. 3.1 (a) American Common Goldfish. (b) European Common Goldfish.

margins) appreciably shorter than the body length. The true identity of this form is probably the Japanese Shubunkin, which Wolf does not mention by name. He does mention specifically Comets with deep ox-blood coloured bodies, white, elongated fins, widely spread full or bifurcated tail fins carried horizontally and as long or longer than the body. Characteristically the Comet is capable of rapid movement.

3 *Japanese Fringetail Goldfish (Fig. 3.3)*

To many in the West this variety is the epitome of what a fancy goldfish should be. The name 'Fringetail' is quite inappropriate as it would, if interpreted literally, suggest a fish in poor condition with frayed finnage. This name has been superseded by the designation 'Veiltail' which is generally used in the English-speaking world, largely because it is so very appropriate. The published drawing by the author has long been the classic depiction of the variety, about which he is very enthusiastic, an enthusiasm which has been sustained throughout

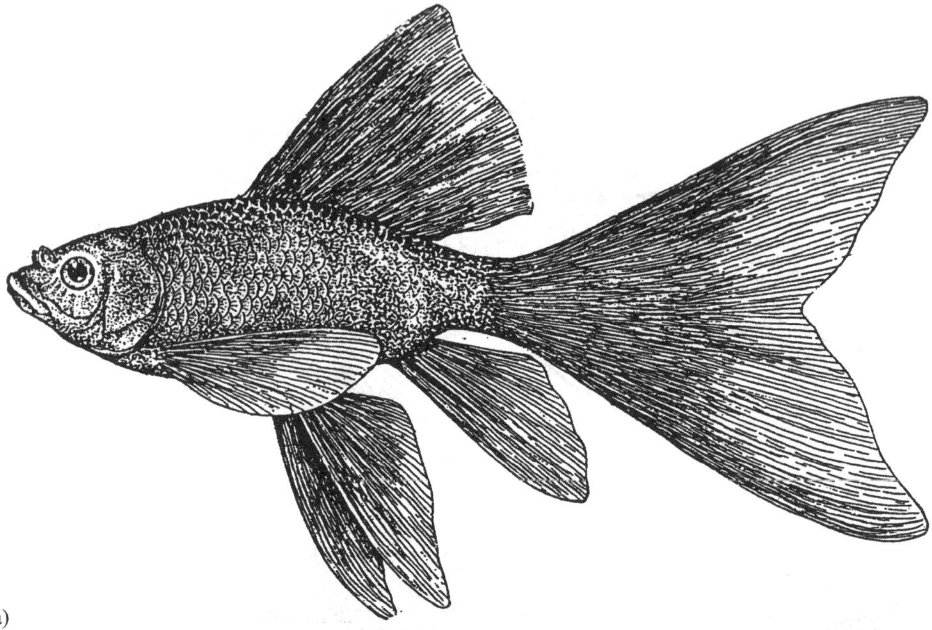

(a)

(b)

Fig. 3.2 (a) Metallic-scaled Comet. (b) Transparent-scaled Comet (possibly original type Shubunkin). (Wolf 1908.)

the twentieth century by those who in Wolf's words regard it as 'the handsomest of all goldfishes'. This enthusiasm is remarkable as in neither Japan nor China is it very highly regarded. The specimens sent to the Chicago World's Fair in 1893, which were later rescued and nurtured, appear to have been make-weights and not specimens particularly prized by the Japanese themselves.

The salient features of Wolf's description are that it has a short body and head, an almost egg-shaped body with all paired fins long and pendent, the dorsal he describes as long, wavy and lace-like (at the present time it would be described as high or tall and sail-like). The

Fig. 3.3 Japanese Fringetail or Veiltail (Wolf 1908).

caudals are described as 'immense, delicate and drooping, divided to the base...much longer than the body of the fish' (in the drawing it falls only a little short of being twice the body length). Features depicted in the drawing which are of current interest are the body proportions in which body length is about one and a half times the depth. The dorsal and ventral contours are both smooth, nearly symmetrical curves. The carriage of the caudals continues broadly the curvature of the back with a slight change at the caudal peduncle. The height of the dorsal is approximately equal to the depth of the body. Wolf refers to its rich, burnished metallic lustre, which can also be developed very impressively in modern examples. The example given by Wolf still represents the ideal Veiltail; in only one detail has there been a change of note. While Wolf's fish show detectable bifurcation of the caudals the modern preference is for a more square-cut broad tail. There can be little doubt that a reincarnation of Wolf's fish even today would carry all before it at a goldfish show. An important point made is that it is only in mature fish that the full development of its characteristics occurs.

An interesting point which Wolf makes is that American breeders had crossed Chinese fish with transparent scales with Japanese Fringetails to produce what we would recognise as Calico Veiltails. These are stated by Wolf to be even more colourful and handsomer than the metallics. This presumably was the beginning of the famous Philadelphia Calico Veiltails which have made a lasting impression on British goldfish breeders.

4 The Japanese Fantail

Wolf observed that although the Fantail had points in common with the Fringetail, it nevertheless merited recognition as a distinct variety. He mentioned significantly that there was a tendency for underdeveloped Fringetails to be lumped in with the true Fantail. Thc latter shared a similar body shape with the Fringetail, the length of finnage was shorter; however, the major distinguishing feature was the fact that the caudal peduncle showed no downward curve as in the Fringetail but was horizontal. It is interesting to note that even at the present time sub-standard Veiltails are passed off as Fantails. It is also interesting to note that Wolf states that at the time of writing there were no transparent-scaled Fantails. A similar range of metallic colours occurred as in the Fringetail.

5 The Japanese Nymph

The Nymph is essentially a single-tailed version of the Fringetail. Wolf considered that it could be produced from a cross between a Comet and a Fringetail. In the first half of the past century it was recognised as a variety but in the latter half this has not been the case. It is now considered as a somewhat aberrant form not to be recognised as an exhibition variety. This does not mean that it is not an attractive fish, which it certainly can be and which goldfish keepers might be happy to accommodate. The carriage of the tail may vary depending on its length; it may be held more or less horizontally if not too long while long tails may be inclined to drape like those of a Veiltail. They may be either metallic or calico.

6 The Japanese Hooded or Lion-headed Goldfish (Fig. 3.4)

The Lionhead which Wolf has described has a body and finnage not dissimilar to that of the Fringetail except that the dorsal fin is lacking. Its major distinctive characteristic is the hood-like excrescence enveloping much of the head. Wolf also mentioned Korean strains showing similar features, the 'Ranchiu' and the 'Maruko'. However, he does not mention that these two types both have quite short fins and that the Maruko does not develop a hood. In fact what he described corresponds to descriptions of the 'Shukin' which has been described from Japan but is not commonly seen at the present time and is essentially a long-finned Lionhead.

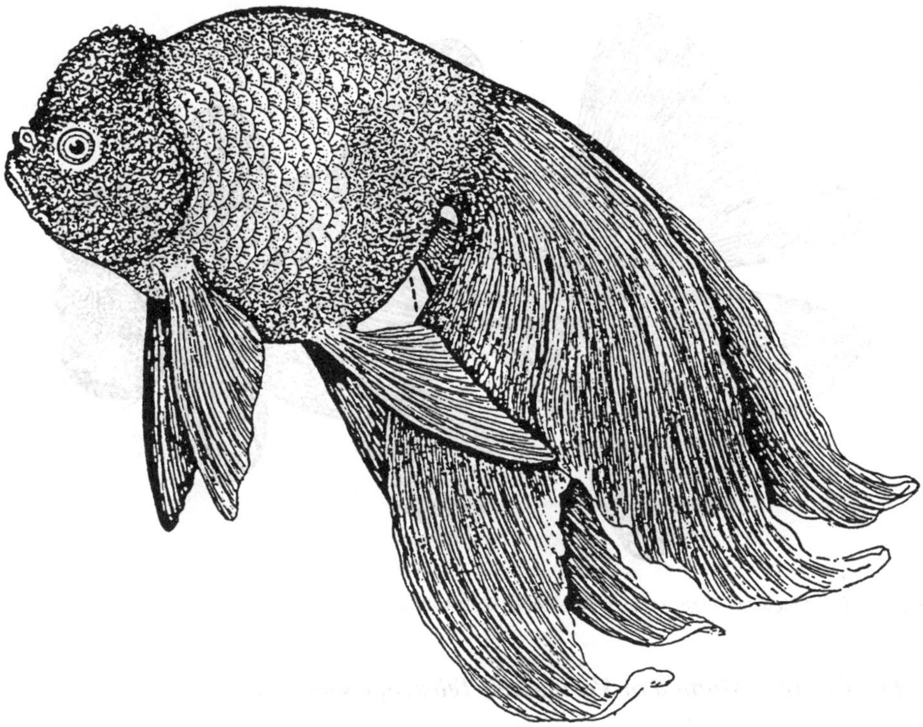

Fig. 3.4 Lionhead – the conformation of this fish is actually that of the shukin (Wolf 1908).

7 The Japanese Barnacled Paradise Goldfish

This was considered an exceedingly rare form with telescope (or globe) eyes and fin development comparable with that of the Fantail but with a square-cut caudal. The rather tubular eyes faced forward. The distinctive feature described by Wolf is that the skin was covered with wart-like growths or papillae…giving the fish somewhat the appearance of being covered with barnacles. Colours recorded are 'mottled red and white with black and white fins and tail'. There is no modern fish which quite corresponds with this description but, the telescope eyes apart, its closest modern counterpart is probably the Pearlscale.

8 The Chinese Telescope Goldfishes (Fig. 3.5)

Effectively these constitute a family of goldfish varieties characterised by the possession of telescope or globe eyes. The basic type can be regarded as like the Fantail in body and finnage (with a rather square-cut caudal). A range of colours is reported, bodies may be red, black or white. Coloration may be very striking indeed with some mottling but Wolf did not consider the coloration he saw as being necessarily fixed or stable. He recognised six different forms of eye development, ranging from the flat, non-protuberant eye of the Common Goldfish through four different types of telescope eye in which the protuberant eyeball shows slightly different forms to the Celestial, which is not only protuberant but upwardly rather than laterally or forwardly directed.

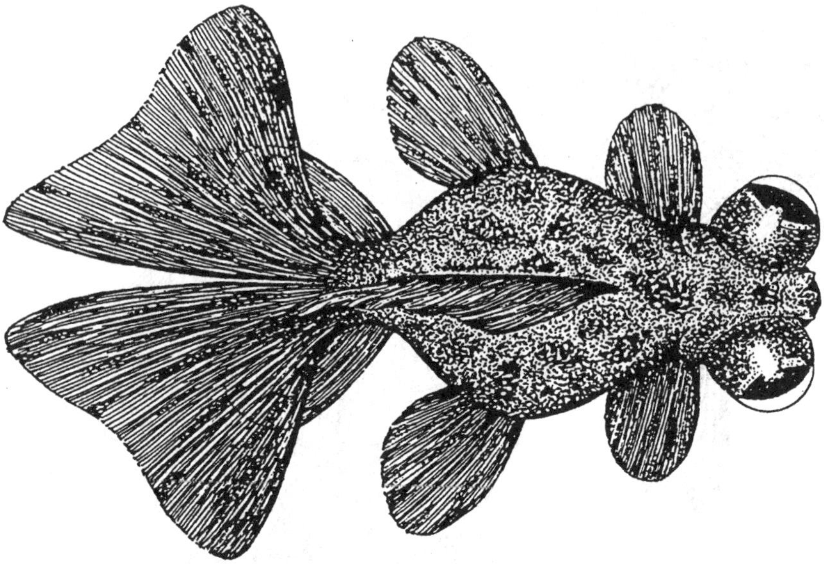

Fig. 3.5 Dorsal view of Chinese Calico Telescope (Wolf 1908).

9 The Chinese Mottled or Variegated Telescope Goldfish

The term 'calico' was probably first applied to this goldfish variety to describe its attractive pattern of coloration. The mixture of blue, red, yellow, orange, white, brown, black and even green is infinitely variable and can in the best examples be absolutely breathtaking. It is probable that this variety is the source of the calico characteristic which has been transferred to the whole range of single and twin-tailed goldfish varieties.

The body form is similar to the Fantail generally but the caudal peduncle has a downward curve. The same applies to the finnage except that the caudal fin is square-cut. Earlier in the century this fish was highly regarded; nowadays it is not commonly seen.

10 The Chinese Fringetail Telescope Goldfish

The fish Wolf described under this name is essentially a Calico Fringetail Telescope. He also described a transparent variant in which the vertebral column and internal organs can be made out. From our genetical knowledge we can infer that individuals of this type are homozygous for the mutation which is expressed as 'calico' in heterozygotes.

11 The Chinese Moor Telescope Goldfish

Although the fish described by Wolf is long-finned, he does not consider it to be a Fringetail, but rather a long-finned version of the Chinese Telescope (8). The chief characteristic which distinguishes it from the other Telescope varieties is the black coloration. Although, as he mentioned, this fish has metallic scales, the density of the black pigment is such that it has the appearance of black velvet. He noted that in predominantly black fish the abdomen might have a yellowish tinge in some individuals; in others yellow was absent. He was of the opinion that the black colour in the latter was more stable. The situation he described has parallels with the current position (AD 2000) in which the Chinese are continuing to develop variations on the telescope theme on a very broad scale. Wolf's reservations about stability of colour and pattern are still valid; notwithstanding this, these variants are both interesting and attractive even if rather ephemeral.

12 The Chinese Piebald or Tiger Telescope Goldfish (Figs 3.6a, 3.6b)

This is a most unusual fish, the body section being triangular and short. Wolf considered it to be deformed. The colour consisted of dark patches of black, brown, red and grey interspersed with lighter areas produced by transparent scales through which pink and blue colours could be seen. Also mentioned are reports from Europe of fish marked with bands of contrasting colours. This is another of those extraordinary goldfish which appears out of the blue from China, makes an impression and then disappears. It may be that such variants for whatever

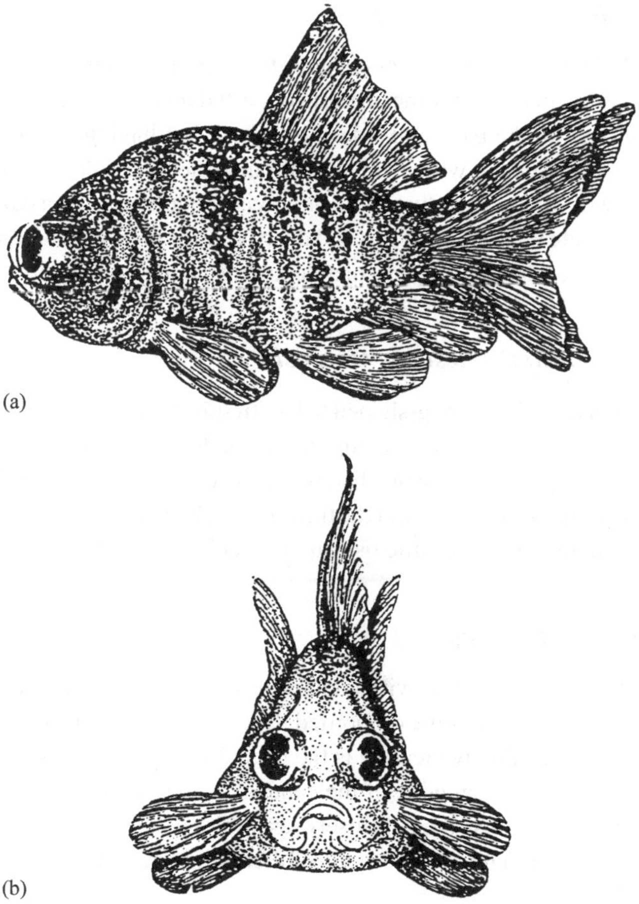

(a)

(b)

Fig. 3.6 Chinese Tiger Telescope: (a) lateral view; (b) head-on view. The colour, patterning and body conformation (especially the triangular body section) are most unusual (Wolf 1908).

reason can only be produced successfully in small and limited numbers and do not lend themselves to continuing commercial exploitation.

13 The Chinese Lettered Telescope Goldfish

In both the eighteenth and nineteenth centuries there are reports and depictions of fish bearing markings which took the form of Chinese characters. The modern view is that these are not natural and were produced by painting the surface of fish with acids, caustic materials or dyes to achieve the desired effect. In recent times there has been an outcry provoked by the sale of imported tropical fish which have been dyed and at least locally the practice is deemed totally unacceptable. There are reports of minor surgery on fishes to remove scales, for example those which mar a colour pattern. Such practices are also highly dubious. The example described is otherwise very similar to the Tiger Telescope and in Wolf's opinion was described in the eighteenth century by de Sauvigny in his *Histoire Naturelle des Dorades de la Chine* (1780) under the name Quen-yu.

14 The Chinese Blue Telescope Fish

This category includes two readily distinguishable types from the published description. The first is described as 'scaled' with a metallic sheen, a 'silvery abdomen flushed with rose-pink, a rich azure blue on the back and sides, the whole fish having a metallic lustre'. The second is 'transparently scaled, with a velvety, ultramarine blue color on the back, reddish blue transparent lower sides and a blue-white or greyish abdomen, with a dark bluish-brown or black dorsal fin, white or grey lower fins and dusty-grey or brownish double tail'. The fish is not illustrated as the author said he was unable to find a perfect example to draw.

15 The Chinese Celestial Telescope Goldfish

This is described as a dorsal-less, egg-shaped fish with spheroidal upturned eyes. The caudal fins are carried horizontally and are of comparable size to those of the Fantail; the caudals show very slight forking. At the time this fish was considered a difficult subject, not easy to keep, and as far as the author was aware at the time had not been bred in the United States. The illustration shows an obviously metallic fish but no specific colours are mentioned.

16 The Chinese Eggfish (Fig. 3.7)

This is a twin-tailed dorsal-less fish with a characteristic egg-shaped body. The scales are metallic and colour may be red, white or mottled. Eyes are normal. The fish was uncommon in the United States. Its distinctive feature is the caudal fins which are long but narrow and drooping, the curve of the dorsal edge of the caudals is continuous with that of the dorsal contour. Among modern goldfish it most closely resembles the Phoenix in general conformation, although the latter has a longer body and a fuller tail, carried in the same fashion.

17 The Chinese Tumbler Goldfish (Fig. 3.8)

Fish of this type were described by de Sauvigny as Kin-teon-yu. Wolf's description and draw-

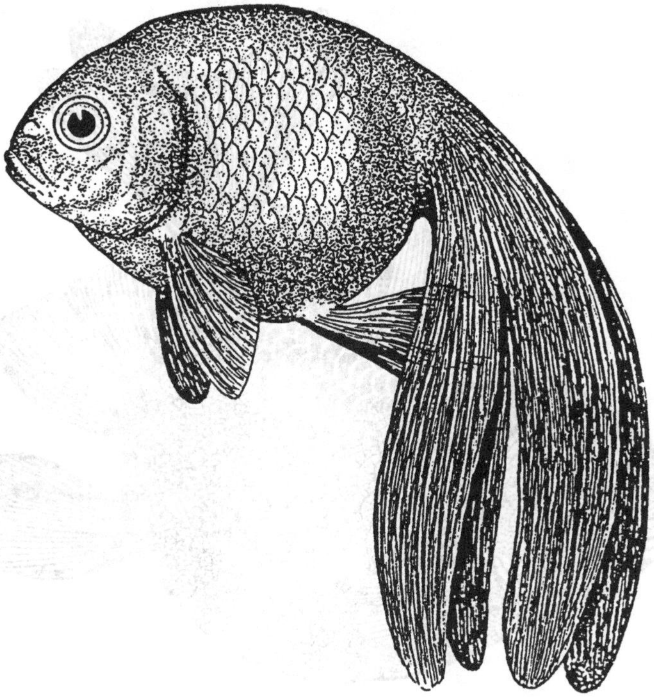

Fig. 3.7 Chinese Eggfish. In comparison with the modern Eggfish the caudal is very much longer. Elongation of the body would produce an approximation to the modern Phoenix (Wolf 1908).

ing is based on information and material supplied by Mr Hugo Mulertt. This fish is unable to swim normally because of the extreme curvature of the spine which gives a markedly concave dorsal and extremely convex ventral contour. It is a metallic telescope with a blue colour flushed with orange. In attempting to swim, the fish tends to somersault in a way reminiscent of the flight of the Tumbler pigeon. This variety, if still in existence, would be unlikely to achieve popularity.

18 Agard's Wonder (Fig. 3.9)

This is a truly extraordinary fish produced by crossing a transparently scaled Comet and a similar Telescope. The resultant hybrid showed the eyes and body form of the Telescope and the scalation common to them both. The twin caudals were longer than the body. The most remarkable feature is the lengthy vertebral column which produces an extremely elongated caudal peduncle. Reputedly the head of the fish could become enveloped in the folds of the tail when at rest.

19 The White Rat (Figs 3.10a, 3.10b)

This unusual fish was named from its appearance when viewed from above. It is a single-tailed Telescope with an egg-shaped body. From Wolf's remarks it is possible that these and similar variants which might arouse disbelief are occasionally depicted on Chinese ceramics. The accuracy with which the form of goldfish is reproduced on art works in general inclines

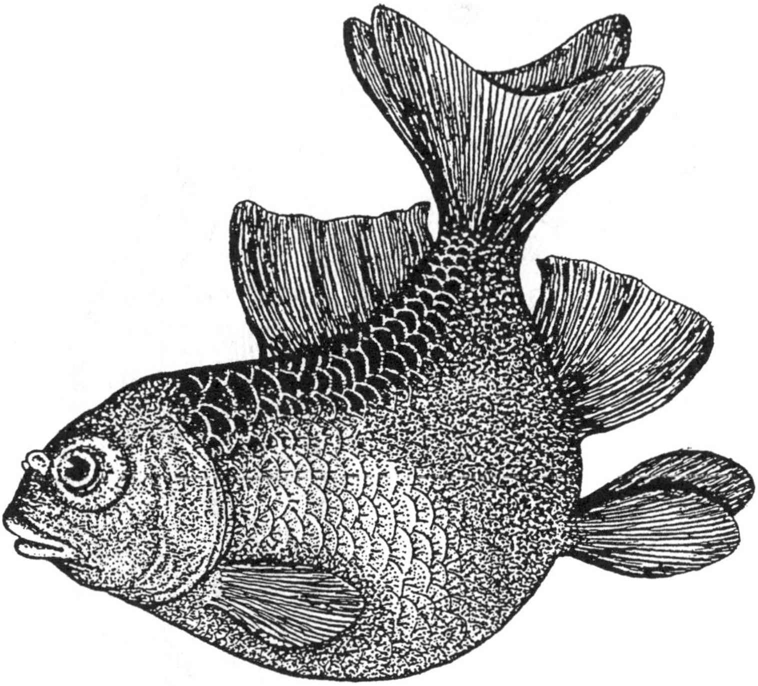

Fig. 3.8 Tumbler goldfish. The curvature of the spine produces a markedly concave dorsal contour; the combination of the angle of the caudal peduncle and the twin tails results in an erratic tumbling motion (Wolf 1908).

Fig. 3.9 Agard's Wonder. The notable features of this variety are the unusually elongated caudal peduncle combined with very long caudal fins (Wolf 1908).

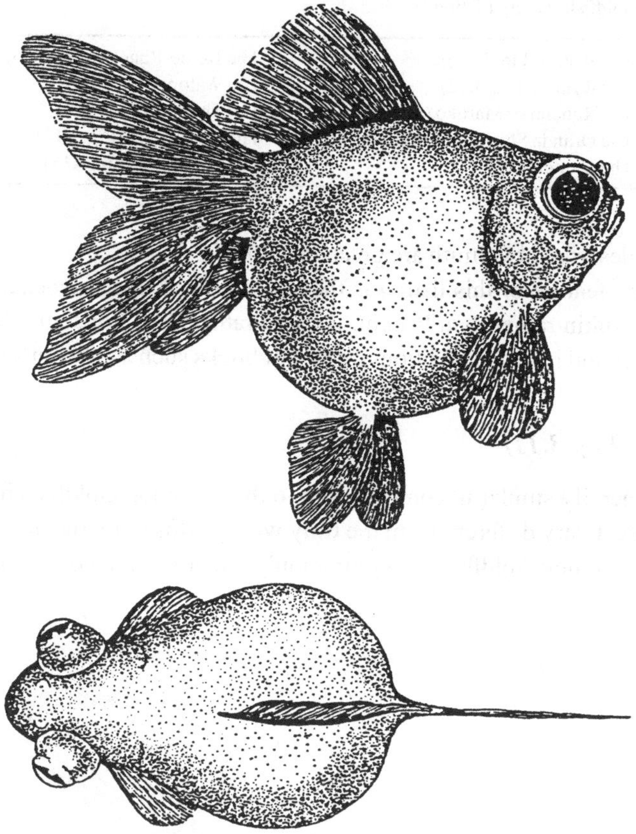

Fig. 3.10 The White Rat: (a) lateral view; (b) dorsal view. This single-tail variety shows an unusually short and egg-shaped body, combined with telescope eyes and loss of melanin and xanthic pigments (Wolf 1908).

one to accept them as truly representative. Figures on ceramics and other materials are invaluable in tracing the historical development of the goldfish over the past millennium. Such works of art can be dated readily. Wolf was prepared to accept them as accurately recording the morphology of the goldfish, an opinion which the passage of time has strongly reinforced.

Hugh M. Smith (1909) *Japanese Goldfish – Their Varieties and Cultivation*

It is remarkable that in the year following the publication of Wolf's *Goldfish Breeds* that another major and original goldfish publication should appear. The background of Wolf's publication was the Aquarium Society of Philadelphia while that of the second landmark publication, Dr Hugh M. Smith's *Japanese Goldfish*, was the United States Bureau of Fisheries in Washington DC. Would that we had a comparable level of official interest at the present time! Dr Smith's publication is devoted in its entirety to the goldfish, both its varieties and its husbandry; we shall be concerned with the varieties: these are of exclusively Japanese origin whereas Wolf's treatment covered numerous Chinese varieties as well.

Table 3.2 Smith's Goldfish Variety Listing (1909).

1 The Wakin (Fig. 3.11)	6 The Deme-Ranchu (Fig. 3.15)
2 The Ryukin (Fig. 3.12)	7 The Watonai
3 The Ranchu or Maruko (Fig. 3.13)	8 The Shukin (Fig. 3.16)
4 The Oranda Shishigashira (Fig. 3.14)	9 The Shubunkin (Fig. 3.17)
5 The Demekin	10 The Kinranshi (Fig. 3.18)

The varieties described by Smith are given in Table 3.2.

The interesting feature of this listing is the modernity of its nomenclature, even to the spellings used: Ryukin rather than Riukin, Ranchu rather than Ranchiu. These names and spellings will be found in current editions (1991) of books such as Matsui's *Goldfish Guide*.

1 The Wakin (Fig. 3.11)

The Wakin is generally similar in conformation to the Common Goldfish, in other words it is not in this respect very different from the truly wild goldfish or Crucian carp. In colour it is similar to the Common Goldfish and is commonly red or variegated red and white, but the

Fig. 3.11 Wakin (eighteenth-century Chinese scroll). Note 'web-tail'.

full range of goldfish colours have been reported by Smith in the Wakin. The distinguishing feature of the Wakin is that both the caudal and anal fins may be completed duplicated. The other fins are very much as they are in the Common Goldfish. In size the individual caudal and anal fins are the same size as their single counterparts in the Common Goldfish. The ultimate Chinese origin of the Wakin is recognised, from which the Japanese stock was obtained. It was also known at that time in Europe and America. Smith noted the observation of Professor Watase that the duplication of the fins is accompanied by duplication of the supporting bony structures. Duplication of fins may be incomplete resulting, for example, in fins in which the lower lobes are duplicated but not the upper, commonly called tri-tails. Duplication may be almost complete with only the dorsal edges of the paired caudal fins fused, such individuals are called 'web-tails'. The spontaneous duplication of fins in this manner appears to be a unique development in the goldfish. The Wakin is capable of growing to a large size, commonly of 6–10 inches but exceptionally up to 16 inches in length.

2 The Ryukin (Fig. 3.12)

This is the Japanese name for what has been known in English as the Fringetail. It is thought to have been developed from the Wakin by long-sustained selection in Japan. The name comes from the Ryukyu Islands lying between Formosa (Taiwan) and Japan; the basis for this association is that these were a possible route of entry to Japan.

Characteristic features of this variety are the greatly shortened and deep body, a full, rounded abdomen and long flowing fins. The caudal fin is deeply indented and may be as long or even longer than the body. The strong curvature of the spine plus the extensive development of the caudal fins frequently conceals the anal fins. The back is rather humped and the lateral line describes a double curve, which gives some reflection of the modified processes of growth and development which produce the shortened body. It is as though growth in body

Fig. 3.12 Ryukin (Smith 1909).

depth is not inhibited while that in length is strongly curtailed. The impression of body depth is considerably enhanced by the height of the dorsal fin and the length of the pelvics. Smith made the very interesting observation which goes to the heart of the appeal of this variety, which is its graceful and dignified motion. The most active movement is brought about by a rapid flurry of the tail region and caudal fins while gentle movement is effected by means of the motion of the large pectoral fins which produce an elegant and measured movement. This has been likened to the ladies at the Japanese Imperial Court in former times, walking sedately with grace and dignity in their long elegant robes. Characteristics such as these are part and parcel of the appeal of the more exotic varieties.

The Ryukin is remarkably hardy considering the extent of the change which has come about during the course of selection under domestication. The history of this variety during the twentieth century, especially in the Far East, shows that end points in progress of selection have still not been reached, in some cases at least.

3 The Ranchu (Fig. 3.13)

Smith adopted the current Roman spelling which superseded the earlier 'Ranchiu' and 'Ran-tyu'. He has given synonyms of 'Maruko', 'Shishigashira' and 'Korean Goldfish', the latter signifying the route by which this variety perhaps entered Japan. As defined the Ranchu has a short, rounded body, a broad head, short twin caudal fins and paired anals. The other fins are also short. The caudal peduncle is short though thick and the body cross-section is almost circular; overall it tends to the globular, almost egg-shape. In Smith's view, the development of the hood is not necessarily the defining character of the Ranchu. As currently understood the 'Maruko' does not have a hood while the 'Shishigashira' by definition does. Latterly the name 'Maruko' is assigned to the 'Eggfish' category from which it differs mainly in the form

Fig. 3.13 Ranchu, Maruko or Lionhead (Smith 1909).

of the skull, which is closer to the Lionhead or Shishigashira type. Notice was taken of the different extents to which the hood could develop and its colour could be quite variable, as could that of the fish as a whole. Initially the colour seems to have been self-coloured reddish with a bright red head. But as has been observed repeatedly in the goldfish, the colour breaks up and variegated patterns are produced.

Smith considered the Ranchu as a clumsy mover; its motion can be considered as a kind of aquatic waddle. This is certainly not without its appeal and while he does not mention temperament, the combination of the characteristic gait and a friendly disposition is possibly an extremely potent factor in the longstanding popularity of this variety.

The nature of the hood has been explored; it is soft to the touch and represents the enlargement of the normal papillae on the head and is entirely non-malignant. The texture of the hood itself may vary ranging from what looks like a collection of warts (or the fruitlets of a raspberry or blackberry) to a fine-textured foam-like mass.

4 The Oranda Shishigashira (Fig. 3.14)

The name in Japanese means literally Dutch Lionhead. The name does not imply that it came from Holland but that it is in a way strange or outlandish, in much the same way that we use the term 'double-Dutch'. Reputedly it was first produced in about 1840 by hybridisation of the Ranchu and Ryukin, and from its appearance it seems to combine the hood of the Ranchu (i.e. Lionhead) with the body and finnage of the Ryukin. It is noted that the depth of the body is less than in the Ryukin and more elongated than the Ranchu. Two major forms of hood are recognised, one restricted to the upper cranial surface (goosehead), the other enveloping both the upper and lateral surfaces of the head.

Fig. 3.14 Oranda Shishigashira (Dutch Lionhead – Smith 1909).

Apparently the original stock was self-coloured red; variegated red and white forms appeared quite soon afterwards. Uniformly velvety-black forms have been developed. Smith observed an outstanding male Oranda with 'a red head, a yellow-golden body, a black back and black fins'. It is quite apparent that features which are characteristic of the modern Oranda, namely its ability to produce interesting and colourful variants as well as its capacity to grow to a large size, were manifest over 90 years ago. The Oranda was considered to be intermediate in hardiness between its two reputed parents.

5 The Demekin

The Japanese name 'Demekin' means literally 'pop-eye goldfish' which is a totally appropriate name. The name 'Telescope-fish' was not considered appropriate by Smith nor by many others since. The name 'Globe-eye' was coined by the Goldfish Society of Great Britain (GSGB) but the name 'Telescope' seems to be so deeply entrenched as to be incapable of being supplemented by any more rational substitute. My own inclination is not to be over-pedantic in the matter. The Chinese call this variety 'Dragon-eye', a term which Chinese goldfish exporters still favour at the present time. Like so many goldfish varieties it was originally developed in China and rather surprisingly was unknown in Japan until the closing years of the nineteenth century. The Japanese accepted it with enthusiasm and produced a range of new variants by selective breeding in only a decade or so by Smith's reckoning.

The pattern of development of protuberant eyes was recorded; initially the eyes appear normal but progressively they enlarge relative to the size of the body. The size of the eyeball and extent of the protrusion is variable.

The fish as introduced from China had a short, deep body with rather short fins, comparable to those of the Fantail. The achievement of the Japanese was to select for increased length of all the fins. Brilliant coloration was not a feature of the Demekin in Smith's experience; uniform black coloration was common otherwise coloration was pale red or yellowish combined with small black spots or irregular black patches. Fins could be red or black or black with pale red or orange bases. In some Chinese fish 'three or four irregularly distributed or mottled colours in a single fish' were recorded; in all probability this was similar to Wolf's Calico Telescope.

It is interesting to note that at the present time most of the commercial production is close to the original Chinese type and predominantly black in colour, appropriately called by some fanciers 'Fantail Moor'. The term 'Moor' is generally used to denote the black Demekin, which neatly circumvents the problematic name 'Telescope'.

The abnormal form of eye seriously impairs the vision of the Moor and perhaps accounts for the solitary disposition which Smith observed. Goldfish are gregarious by nature and their propensity to shoal can be witnessed in any large pool where they occur in numbers. The poor vision of the Demekin and the eye protrusion renders the eyes susceptible to damage and total blindness.

6 The Deme-Ranchu (Fig. 3.15)

Literally the Japanese name means 'pop-eye Ranchu' implying both protuberant eyes and lack of the dorsal fin. The figure of Smith illustrates a dorsal-less fish with telescope eyes

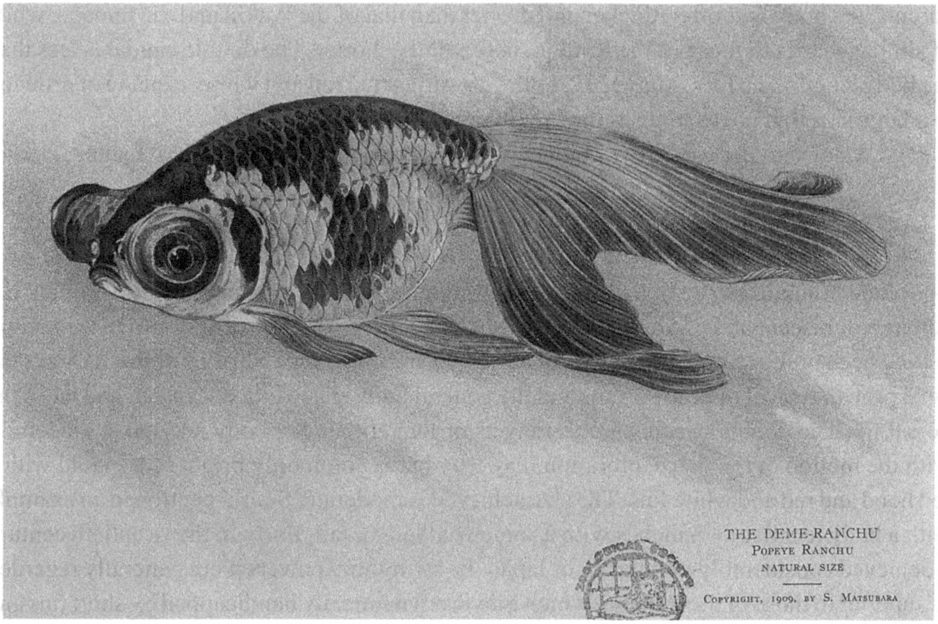

Fig. 3.15 Deme Ranchu (Smith 1909).

similar to those of the Demekin. In his text, however, he describes the unusual orientation of the eyes in which the pupils are directed skywards. The fish depicted appears to represent an intermediate stage between the Demekin and the Celestial proper, which he describes in some detail. The Celestial is of Chinese origin and was taken to Japan comparatively late in the very early years of the twentieth century, appreciably later in fact than its introduction to the Western world.

In body shape the Deme-Ranchu is similar to the Ranchu proper, that is with a short, egg-shaped body; the skull also is broad. The size of the eye is very large and the extent of protrusion is variable. When the Celestial trait is fully expressed both eyes are directed vertically upward; however, it is not uncommon for this characteristic to be unequally developed. The finnage is appreciably longer than in the Ranchu with widespread caudals and in length equalling or exceeding that of the body.

Like the Demekin, the Deme-Ranchu was reputed to be a solitary fish with poor vision and a strong liability for the eyeballs to be damaged by accidents with resulting blindness. Vitality was considered to be low and the ability to reproduce impaired. All these factors contribute to a low level of popularity.

7 *The Watonai*

Smith stated that according to Professor Matsubara the name means 'a variety found neither in Japan nor China'. The variety apparently arose naturally in a pond containing a mixture of goldfish varieties including the Wakin and Ryukin which are believed to be the parents of this hybrid. This is thought to have occurred in Tokyo around 1880 and the variety was first shown to the public in 1883. The body form is intermediate between those of its presumed parents; the body is shorter, thicker and deeper than that of the Wakin and the finnage, while of similar shapes to those of the Ryukin, is generally shorter. The double caudal is less than the body length. The form known to Smith was variegated red and white, capable of growing to a large size (but not as large as the Wakin) and very hardy.

The Watonai as described by Smith corresponds closely with the modern Fantail.

8 The Shukin (Fig. 3.16)

The name 'Shukin' can be translated as the 'Autumn Goldfish' on account of the brilliant red colour reminiscent of the autumn colour of the Japanese maple. The variety is of hybrid origin having been produced from the cross of a Ranchu and an Oranda Shishigashira. It shows the head growth typical of both parents and the general body shape of the Ranchu, and lacks the dorsal fin. The double caudals are as long as or longer than the body and move gracefully with the motion of the fish. Coloration may vary but is commonly bright red or gold with a red hood and red and white fins. The size achieved is moderate. Smith mentioned an example with a body length of 4½ inches with a somewhat longer tail. Early in the twentieth century it achieved considerably popularity in Japan. Its swimming powers were generally regarded as superior to those of the Ranchu, which is hydrodynamically handicapped by short fins and the lack of the dorsal fin.

9 The Shubunkin (Fig. 3.17)

This variety is the only single-tailed fish to be considered by Smith and it has probably achieved its greatest popularity in the West. The Japanese name, awkwardly translated as

Fig. 3.16 Shukin (Smith 1909).

Fig. 3.17 Shubunkin (Smith 1909).

'vermilion red dappled with different hues' is very appropriate and descriptive and has been adopted without dissent in the English-speaking world. This is unusual to say the least.

The Shubunkin is very much a fish of the twentieth century, originating in 1900 according to Smith on information supplied by Professor Matsubara from crosses between the Wakin and Chinese Calico Demekin. Both parental strains were variegated, the Wakin parents were variegated with red, black, bluish and white while the Demekin had 'black dapples or vermilion or purplish body'. This cross produced a mixed F, some of which 'had the form of the wild goldfish and the peculiar markings of the demekin, some resembled the Wakin; some had the form of the demekin'. About 20% were of the desired type and these were selected and bred and gave rise to 'new and interesting color phases'.

The early Shubunkin had a rather elongated and compressed body, with definitely curved dorsal and ventral contours, a very distinct caudal peduncle, inconspicuous scales and somewhat elongated finnage. The caudal is bilobed, i.e. single deeply forked and 'three-fifths to two-thirds the length of the body', that is substantially longer than that of the Wakin parent.

The characteristic mottled colours of the Shubunkin are its chief attraction and from earliest times have been the subject of strong selection. Ideally there is an overlying pattern of dark spots on a mottled background comprising red (vermilion), white, bluish, purple and other colours. Bright red fish with black spots and 'uniformly purple' individuals were produced. The progeny produced variants which according to Smith were 'quite unknown in the parent stock on either side'.

10 The Kinranshi (Fig. 3.18)

The Kinranshi, translated as the 'Brocaded Goldfish', was the product of a Japanese breeder Akiyama Kichigoro about a century ago. The parents of this cross were Ryukins and Ranchus and the result was rather surprisingly a dorsal-less fish with an elongated body, a gently arched back, small fins and a double caudal. The colours developed are red, black and white in varying proportions. This variety has never become popular but it is noted by Matsui in his book (1972).

Fig. 3.18 Kinranshi (Smith 1909).

W. T. Innes (1917–1932, 1947) *Goldfish Varieties*

Innes' work on goldfish varieties, especially between the wars, became a classic and standard work on the goldfish. The inter-war period saw fifteen revisions with two reprintings after the end of World War II. This valuable work had clear antecedents in Wolf's pioneer studies, many of whose drawings have been used in the original or slightly modified and redrawn. Apart from the exclusion of coverage on tropical fishes and its substitution by a treatment on water lilies and water gardens, the 1947 printing is not substantially different from that in the 1932 edition. It is clear from some of the author's prefaces that he considers the position regarding goldfish varieties to be relatively stable. It is interesting that the detailed descriptions of Wolf and Smith are replaced by very concise descriptions, thirteen goldfish varieties are described in thirteen pages of text (including drawings) and six pages of photographs. The varieties described by Innes are listed in Table 3.3.

Table 3.3 Listing of Goldfish Varieties Described in Innes' *Goldfish Varieties* (1917 *et seq.*).

1 The Common Goldfish	8 The Chinese Scaleless Telescope Goldfish
2 The Comet Goldfish	9 The Scaleless Veiltail Telescope
3 The Shubunkin	10 The Chinese Moor Telescope Goldfish
4 The Fantail	11 The Chinese Celestial Telescope Goldfish
5 The Japanese Fringetail Goldfish	12 The Lionhead Goldfish
6 The Japanese Nymph Goldfish	13 The Oranda
7 The Chinese Telescope Goldfish	

1 *The Common Goldfish*

The fish illustrated by Innes is similar to Wolf's 'Common American Goldfish' with an elongated body flattened on the sides. The head is without scales, relatively short and wide. The paired fins are the pectorals and pelvics, the dorsal, caudal and anal fins are all single. Innes does not indicate the range of colour phases in this description but mentions elsewhere red, white, olive-grey and black in addition to red and white variegated as being found in goldfish varieties with metallic or reflective scales.

2 *The Comet Goldfish*

The origin of the Comet is given as the ponds at the Fish Commission in Washington and first exploited commercially by Mr Hugo Mullertt (1883). The body form, while similar to that of the Common Goldfish, is more slender. Its most remarkable feature is the elongation of all the fins. The caudal and anal fins are single and the length of the caudal, measured along either the dorsal or ventral margin, is equal to or greater than the length of the body. It is considered to be an ideal pond fish as it is hardy, graceful and a rapid swimmer. It also breeds reasonably true to type.

3 *The Shubunkin*

The last of the single-tail breeds described is the Shubunkin which Innes considers to be simply a transparently scaled, highly mottled Common Goldfish. This description is certainly applicable to its body form but the fins are generally longer, the length of the caudal is actually intermediate in length between that of the Common Goldfish and the Comet. The Japanese, between the wars and subsequently, bred them in quantity for export. They are equally suited to aquarium and pond, showing striking variations in colour and pattern. The most highly prized have a blue background, sprinkled with brown and black dots and mottled with dark red and yellow. This coloration and pattern is desirable in all 'scaleless', i.e. transparently scaled, varieties.

4 *The Fantail*

This variety is the first of the twin-tail (double-tail) varieties described. The description given is of a comparatively elongated fish though relatively shorter in the body than the Common Goldfish. The fins are not greatly elongated but appreciably longer than those of the latter. It is considered hardy and agile, well able to cope with pond conditions. Between the wars it was produced in quantity in Japan, America and Europe.

5 *The Japanese Fringetail Goldfish*

The illustration published is the epitome of what the ideal Fringetail should be. It is of the Veiltail or Broadtail type and it is the Veiltail name which has stuck. This serves to distinguish it from other variants of the Fringetail which differ appreciably from it in appearance.

The history of the Veiltail is chequered to say the least. It was thought that in 1893 the Japanese sent a collection of what proved subsequently to be outstanding goldfish to the Chicago World's Fair. Mortality was high both in their journey and subsequently at the exhibition. The survivors were in poor condition when they came into the possession of Mr William P. Seal, who restored them to good health, and one then came into the possession of a Mr Barrett. At that time the fish was not particularly remarkable but with the passage of time it developed the characteristics which made it renowned as the 'World's Fair Fish'.

Recently Mr Joe Lightcap (2000) of the Goldfish Society of America has set the record straight. He quotes from the A*quarium Magazine* of March 1968 to the effect that the Japanese exhibit at the Chicago World's Fair of 1893 was not of live fish but of preserved specimens. The fish which came into the possession of Mr Seal had been imported from Japan by the Wisconsin State Fish Commission for its own exhibit at the fair but were not shown as they were in poor condition.

The Aquarium Society of Philadelphia rightly regarded this fish as epitomising perfection and had a drawing made of it, used it as a society emblem and later had a medal struck bearing its likeness. Other aquarium societies have, I know, followed suit and adopted the Veiltail as a badge motif.

The body of the Veiltail is short with nearly symmetrical and smooth dorsal and ventral contours, the abdomen is full and the body somewhat egg-shaped. The dorsal fin is high, equalling the depth of the body, ideally held erect giving a sail-like effect. All other fins are paired, long and hanging gracefully with a lace-like texture. The eyes are normal and do not protrude. The deportment of the best fish is superb as they move slowly and majestically through the water. The original metallic Veiltail stock has been crossed with the Chinese Scaleless Telescope and by appropriate selection and further crosses produced a Scaleless Veiltail, more commonly called the Calico Veiltail.

6 The Japanese Nymph Goldfish

In essence the Nymph is a single-tailed variant of the Veiltail; the anal fin is also single. Apart from this the other features are as in the Veiltail. An interesting point is that the carriage of the caudal is depicted as horizontal in the drawing even though its length slightly exceeds that of the body. Nymphs are not deliberately bred but may be retained by breeders from Veiltail spawnings as they are aesthetically attractive.

7 The Chinese Telescope Goldfish

The conformation is generally quite close to that of the Fantail with the added feature of protuberant eyes. Eyes may vary in direction; most, however, are laterally directed. They may, however, differ in shape – spherical or conical and more rarely ovoid. They may occasionally be directed forward. The telescope eye develops gradually from the age of two months onward; however, not all progeny of Telescope spawnings necessarily develop the characteristic eyes. Innes makes a very telling point that reduction in size of the eyes can come about when selection is practised for increased length of finnage and shorter bodies, and he has noted deterioration over time in the quality of eyes as compared with original stock. He made no mention of colour so presumably the general colour range of metallic fish occurred.

8 The Chinese Scaleless Telescope Goldfish

Innes emphasised that the term 'scaleless' is a misnomer, but uses it because of its wide currency at the time. He made a very important and significant distinction that 'Plain Scaleless' fish show similar colours to those of metallic fish, red and white for example, but lack their characteristic lustre. The quality of the colour may differ and a very attractive ox-blood shade is recorded. The other possibility is 'Calico' in which Innes said that 'Red, yellow, brown, grey, black, blue and lavender are laid in fantastic blotches and spots over the body, usually on a lighter background. Many small dots of black are sprinkled over the body and fins.' As in the Shubunkin, the most highly prized colour is blue, the greater its extent the more valued the fish. As Innes also said 'The colors seem as though they had been laid on by the delicate hand of a water-color artist.'

9 The Scaleless Veiltail Telescope

This type is an American production; to quote Innes, 'We crossed Japanese Fringetails with Scaleless Chinese Telescopes, thereby producing two new varieties which have become permanent – Scaleless Japanese Fringetails and Scaleless Veiltail Telescopes. Both have been bred for broad-tail qualities (Veiltail), and may be considered an American variation.' Coloration is as in the Chinese Scaleless Telescope and conformation is as in the metallic Japanese Fringetail (Veiltail).

10 The Chinese Moor Telescope Goldfish

The example depicted was drawn from a five-year-old specimen in which the finnage might be expected in later life to become more elongated and Veiltail-like, corresponding very closely to the modern Broadtail Moor. A good specimen of a Moor is a striking fish indeed, a wonderful contrast with colourful metallics and calicos. The black pigmentation may be so intense as to look velvety, obscuring the intrinsic reflectivity of the scales; it is all pervasive except for the lower abdomen where it may shade into a blue-grey or golden-brown. In the latter case there is the possibility that the fish may change colour to red. In the former instance the blue-grey abdominal tint is no guarantee that no colour change will occur. Innes also recorded the experience of breeders that the highest proportion of black progeny result not from black × black matings but from black × red telescope combinations.

11 The Chinese Celestial Telescope Goldfish

Innes considered the Celestial to be the most difficult of all goldfish varieties to keep and breed. He recorded the pattern of development of the eyes as consisting of two phases. In the first the eyes become protuberant as in other telescope varieties, in the second the pupils of the eyes 'gradually turned towards the top of the head'. The peculiarities of the eyes were definitely not produced by any kind of environmental manipulation nor by any mechanical means or contrivance. In any event had such means been effective the resulting change would not have been heritable. The fish depicted in the drawing appears to have a short, rather egg-shaped body with caudals approximately equal to the body in length. The preoccupation with the upturned eyes was obviously so considerable that little detail about other aspects was given.

12 The Lionhead Goldfish

Lionhead is a literal translation of the Japanese 'Shishigashira' and seems to be well entrenched in spite of Innes' own expressed reservations about the suitability of this name. He believed that the name 'Buffalo-head' would be much more descriptive and appropriate. Similar reservations were felt in Great Britain where the GSGB proposed the name 'Bramble head'; certainly there is a close resemblance of the bramble fruit to the hood of this fish, closer in fact than to the manes of either lion or buffalo. Perversely perhaps the name Lionhead sticks except insofar as in recent times it has a rival in the name 'Ranchu', especially as regards Japanese fish.

The dominant features are first of all the growth over the head and gill-plates, resembling (as Innes has stated) a large raspberry (? bramble head!); the second is the total lack of dorsal fin; the third is the very thick short body carrying short fins. The hood growth was thought to impede somewhat movement of the gill-plates, necessitating care in ensuring a good oxygen supply especially in warm weather. As in all twin-tail fish both caudals and anals should be double; interestingly Innes suggested leniency towards deficiency in these features if head and body were good. The customary colours recorded were red and white as in other metallic goldfish; pearl white bodies and pale yellow heads were also noted. Transparent scaled Lionheads have been produced by crossing but not apparently in quantity. Innes dismissed the idea that dorsal-less fish could be produced by extraction of the fin in young fish on the grounds that a change induced in this way would not be heritable.

13 The Oranda

In Innes' opinion the Oranda was a sport from the Lionhead, one which did not reproduce entirely the parental form. He recorded that fins and body are longer than in the Lionhead. It is interesting to note that the Oranda has received the shortest description (four lines) as compared with the Nymph with one of double the length (nine lines)! This is particularly interesting in view of the current popularity (AD 2000) of the Oranda and the complete, total and absolute eclipse of the Nymph! This has been due largely to the propensity of the Oranda to generate a multiplicity of variants, which came to light in the latter part of the twentieth century, and the great popularity these have achieved. The status of the Nymph was called into question soon after World War II; since nobody actually bred the Nymph as a variety in its own right there seemed little justification for its recognition as a variety. With publication of postwar goldfish show standards it was dropped.

Hodge & Derham (1926) *Goldfish Culture for Amateurs*

There can be little doubt that Hodge and Derham were greatly influenced by the previous work of W. T. Innes in producing their own work to which they refer. Varieties listed are in Table 3.4.

This treatment, while covering exactly the same ground as Innes, differs in its treatment of the Telescopes, recognising three types where Innes recognises five. Both scaled and scaleless Chinese Telescopes are considered together, but not the Scaleless Veiltail Telescope and the Chinese Moor Telescope. Much of the material presented has been gleaned from Innes. There are a number of noteworthy points which they raise that are not mentioned by Innes; they noted scaleless Comets but do not specify colour. It may be that these are red and white transparent rather than calico, in which case they are correctly designated Comets; were they calicos the question could be raised as to whether they should be considered as Shubunkins. In considering the Shubunkin, Hodge and Derham mentioned a long-finned strain produced by American breeders and it seems reasonable to conclude that the transparent scaled Comets are truly Comets and not long-finned Shubunkins. The name 'Ryukin' is mentioned in connection with the Fringetail. Interesting if brief reference is made to other varieties, the Chinese Tumbler Goldfish, the Chinese Eggfish, the Meteor Goldfish and the purported variety marked with Chinese characters. The Tumbler, Eggfish and the Chinese Character Goldfish are presumably the same as those mentioned by Smith (1909). The reference to the Meteor is interesting; this variety lacks a caudal fin completely but has compensatory development of another fin (presumably the anal) which assists it in movement. Hodge and Derham indicate that the source of information about this variety is Mr W. T. Innes, this information is not given in later editions of *Goldfish Varieties* but could have been included in earlier editions or articles in *The Aquarium* magazine.

Hugh M. Smith (1924) *Goldfish and Their Cultivation in America*

Dr Hugh M. Smith produced in 1924 an interesting article on goldfish culture in America which is by way of being complementary to his earlier work on Japanese goldfish culture (1909). His listing of goldfish varieties cultivated in America is given in Table 3.5.

There is nothing to be gained by reiterating descriptions of all these varieties which have been given by Smith himself and others. However, there are some points on which comment is appropriate. Applying the name 'Wakin' to the single-tail Common Goldfish is inappropriate and misleading. In Japan the Wakin is actually a twin-tailed fish, the single-tailed equivalent

Table 3.4 Listing of Goldfish Varieties in Hodge & Derham (1926).

1 The Common Goldfish	7 The Fringetail Telescope
2 The Comet	8 The Celestial Telescope
3 The Fantail	9 The Lionhead
4 The Nymph	10 The Oranda
5 The Fringetail	11 The Shubunkin
6 The Chinese Telescope	

Table 3.5 Listing of Goldfish Varieties by H. M. Smith (1924).

1 Common Goldfish – Wakin	8 Chinese Telescope Goldfish
2 Comet Goldfish	9 Celestial Telescope Goldfish
3 Fringetail Goldfish – Ryukin	10 Veiltail Moor Telescope Goldfish
4 Veiltail Goldfish	11 Japanese Lionhead Goldfish
5 Nymph Goldfish	12 Oranda – Dutch Lionhead Goldfish
6 Fantail Goldfish	13 Shubunkin – Calico Goldfish
7 Veiltail or Broadtail Telescope Goldfish	

being the Hibuna or red goldfish. The confusion has arisen probably because the Wakin has been the most popular form of goldfish in Japan. There are some very interesting comments on the Fringetail or Ryukin from which the Veiltail was derived by American breeders. The form of tail fin provides the basis for recognising two distinct groups of Fringetails: those with deeply bifurcated fins are known as Ribbontails or Swallowtails (both terms are to be found in later literature), those with less indentation are Broadtails or Veiltails and both of these terms too have entered the literature. Although not mentioned by Smith, the main distinction between what are currently called Ryukins and Veiltails is the body shape. The Ryukin has a tendency to develop a hump-back while the dorsal contour of the Veiltail describes a smooth curve in the best specimens. Some individuals may show a slight tendency to produce a hump, harking back to their origin.

There are some interesting remarks on the Lionhead, which has a multiplicity of names, and Smith has suggested adoption of the name 'Ranchu' for the Japanese Lionhead. The illustration depicts a fish which would be regarded at the present time as a Ranchu with the highly characteristic strongly curved back and the horizontal carriage and characteristic shape of the caudals. This distinguishes the Japanese from the Chinese Lionheads as portrayed by Innes in *Goldfish Varieties*.

T. C. Roughley (1936) *The Cult of the Goldfish*

Like Dr Smith, T. C. Roughley was a professional biologist. He worked in Australia and produced one of the best works on goldfish in the inter-war years. His listing of varieties (Table 3.6) is rather similar to that of Smith in the previous decade.

This listing is generally similar to that of Smith, even to repetition of the confusion of Common Goldfish and Wakin. Roughley considers Veiltail and Fringetail together. The greatest difference in treatment is with regard to the Telescopes: Smith separates varieties on the grounds of finnage development while Roughley is more inclined to distinguish on scale type,

Table 3.6 Listing of Goldfish Varieties by T. C. Roughley (1936).

1 The Common Goldfish or Wakin	7 Telescope
2 The Comet	8 Calico (Telescope)
3 The Shubunkin	9 Moor
4 Fantail	10 Celestial
5 Veiltail or Fringetail	11 Lionhead
6 Nymph	12 Oranda

metallic versus transparent. Roughley recognised the Calico (Telescope) as the source of the transparent scale character. He advanced the view that the Shubunkin arose directly from the Calico by reversion of its progeny to the wild-type body form and eyes. We know that this actually came about through deliberate hybridisation by the Japanese.

The contributions of Chen (1925) and Matsui (1934)

The specialist literature on the goldfish tends to be of two quite distinct types. On the one hand there are texts which cover all aspects of goldfish husbandry and whose coverage of the variety situation is not very detailed. On the other hand there are works such as those of Wolf and Smith published in the first decade of the twentieth century that were very detailed and extremely valuable. These provided basic material on the variety question which many authors used between the wars without adding a great deal to the sum of our knowledge on the matter. In the period 1917–1949 the various editions of Innes' *Goldfish Varieties* seemed to satisfy the general needs of the enthusiast. The situation changed in the 1950s and subsequently when a number of rather more original and innovative texts were published. It is rather ironic that the origins of this postwar renaissance can be traced back to the work of two Oriental scientists who devoted much of their lives to the study of goldfish. The first was Shisan C. Chen who published in 1925 an extraordinarily detailed account of the morphological variation found in the Chinese varieties of goldfish. The second was Yoshiichi Matsui, a Japanese geneticist who published a series of papers on the goldfish and its genetics in 1934. These two contributions are major benchmarks in goldfish literature. Above and beyond this, both men published other work which advanced our knowledge very considerably.

Chen (1925) *Variation in External Characters of Goldfish* Carassius auratus

Chen was a pioneer in the study of goldfish genetics to which he made very significant contributions but his outstanding work was entitled *Variation in External Characters of Goldfish* Carassius auratus. This contains the most extensive coverage of variation in body form, finnage, shape of head and eye, in the nature and development of scales, the operculum and the nares, and in colour. Three-quarters of a century later this is still without rival as the most detailed and comprehensive account of the morphological variation to be found in the whole range of goldfish varieties. While popular writers between the wars readily acknowledged that the range of forms which could be obtained was anything but the full range of what had been developed in China, it was not until 30 years later that many variants mentioned by Chen appeared in the West. Chen does not in this publication give diagnostic descriptions of varieties as such but in considering individual features refers to varieties which manifest them. It is interesting to extract such references to varieties from his paper, list them (Table 3.7) and make comparisons with the listings of other writers.

This listing presents some curious features, the first being that, while Chen went to enormous trouble to gather together all available goldfish variants, he does not cover the Common Goldfish. In his measurement of body proportions of goldfish, wild and domesticated,

Table 3.7 Listing of Goldfish Varieties taken from Chen (1925).

1 Comet	8 Celestial
2 Shubunkin	9 Bubble-eye
3 Wakin	10 Eggfish
4 Ryukin	11 Ranchu
5 Pearl Scale	12 Oranda
6 Telescope-eye (Fig. 3.19)	13 Narial Bouquet
7 Moor	14 Outfolded Operculum (Fig. 3.22)

he reported that no Chinese domesticated material he examined had preserved the original proportions of the wild fish. The only kind which did so was the Wakin which he had obtained from Japan. It is generally accepted that the Wakin effectively differs from the Common Goldfish only in the possession of a double tail. Chen illustrates a fish which has a slightly shorter body than the wild type but which bears rather longer finnage. His appreciation of what the Comet variety is does not agree with that of the Americans at that time in that he gives the name to the single-tail version of the Fringetail, which was then called the Nymph.

In his consideration of transparent scales, he made mention of the Shubunkin but noted that this type of scale was found in other varieties also. The other scale type described is the Pearlscale which was not to become known widely in the West for another quarter century or so. The Telescope-eye (Fig. 3.19) and the Moor depicted are similar as far as body type is concerned but have longer finnage than the wild-type and deeper bodies, but this falls far short of the finnage and body of the Broadtail Moor, for example. In addition to the Moor the illustrations include 'mottled' (calico), together with brown and blue metallics. Varieties lacking a dorsal fin which he considered are the Celestial, the Bubble-eye, Eggfish and Ranchu.

Fig. 3.19 Telescope-eye (Chen 1925).

The position of the Celestial eye is derived from the Telescope by initial forward rotation of the eye through approximately 90° and a subsequent rotation of 90° to place it in a vertical plane. This can be seen in the course of development of this character and it can be confirmed by comparing the anatomical features of both kinds of eye, which Chen has done.

The Bubble-eye (Water Bubble Eye in full) is a variety described by Chen which did not become extensively known in the West until well after the end of World War II. The size of the bubbles had in the interim been enormously increased by selection in Japan, from where they were exported. Chen interestingly mentioned that development of the sacs or bubbles from the lower orbit of the eye can cause a displacement of the eyeball, simulating the appearance of the eye in the Celestial. The treatment of the other dorsal-less groups is very modern, being considered as either Eggfish or Ranchu. The fundamental difference between them is the form of the skull, which in the Eggfish is narrow and pointed while in the Ranchu it is broad and blunt. The Ranchu group is generally, but not invariably, characterised by the development of the typical hood. The extent of this development varies, in some Ranchu types such as the Maruko it is negligible, in others it may be confined to the cranium area, and in yet others it may cover almost the entire head, the cranium, around the eyes and on the gill-plates. The Oranda shares the characteristic hood with the Ranchu and development of the hood shows similar variation. It differs from the Ranchu in body and finnage, which tends to be more Ryukin-like, and it possesses a dorsal fin. The Narial Bouquet (or Pompon) has hypertrophic nares which resemble powder puffs on its nose. Like the Bubble-eye this did not appear in the West in significant numbers until after the end of World War II.

The Outfolded Operculum (Fig. 3.22, see p. 68) has gill covers with an outward curve at the free end, which exposes the gill filaments to a greater or lesser degree. This character is one which has interested Chinese goldfish breeders but is regarded as a deformity in the West, and while it does appear from time to time in Western stocks, is strongly selected against.

Chen also undertook an analysis of the colours developed in metallic strains. He published illustrations of the pigment cells located in their scales and showed clearly that the xanthic forms had no dark pigment cells present on the scales while the blue metallics had no orange-red pigment cells (xanthophores). In the black or melanic forms there was a dense production of melanophores which masked any xanthophores present, while in brown fish both melanophores and xanthophores contributed to the colour. In all the metallic scale types the sheen was produced by cells which contained a 'bluish, prismatic colour', which are the iridophores. In the paler ventral scales the density of pigment cells is very much lower than in darker areas although the size of individual chromatophores is larger. In lighter areas the blue, prismatic tissue is also more in evidence; the concentration of pigment in the black and orange cells is also lower. Deviation from the wild-type colour is produced by variation in the relative frequency of the different types of chromatophore; either or both the red and black cells may be absent. Silver fish are produced when neither are present and the colour seen is due to the presence only of Chen's 'blue-prismatic tissue', i.e. the guanine-containing iridophores. Chen's was the first attempt to explore the anatomical basis of the colour in metallic scaled goldfish. The colour situation in the transparent scaled fish has not been investigated as thoroughly. This is admittedly a more complicated proposition.

The immediate impact of Chen's work on the popular appreciation in the West of extant available goldfish variants was not very significant. As had already been noted there was almost a canon of goldfish varieties available in the West, the dozen or so considered by Innes in the various editions of his work between 1917 and 1949. Although the goldfish was only introduced to Japan a century or so before it was first taken to Europe, its cult was taken up very enthusiastically by the Japanese who made significant additions to the range of variants developed in the first place by the Chinese.

Matsui (1934) *Genetical Studies on Goldfish of Japan*

We are fortunate indeed that the variation in goldfish which was extant in Japan was studied by the pioneer fish geneticist Professor Yoshiichi Matsui at the Imperial Fisheries Institute. Matsui's treatment includes a listing of Japanese material which is an excellent complement to that which was covered by Chen in the previous decade. Matsui's work was published in 1934 when the preliminary phases of what was to become the Sino-Japanese War (1931–1945) had already been enacted. Matsui's listing of goldfish varieties in Japan is found in Table 3.8.

This listing is substantially that given by Matsui with some minor modifications. There are numerous interesting points made by Matsui on various different varieties which will be noted together with some explanatory comments; repetition of detail given elsewhere will be avoided where possible.

1 Hibuna (Western Common Goldfish)

This is the commonest goldfish in the West and the most popular but not so in Japan. Matsui's brief description indicates a basically wild-type body form with rather pale gold coloration.

2 Wakin (Japanese Common Goldfish)

The body form is slender with short fins. Variation was seen in the extent of development in the twin-tail character. Four types are recognised:

(a) Yotsu-wo (four-lobed) i.e. with completely duplicated caudals

Table 3.8 Listing by Matsui of Goldfish Varieties in Japan (1934).

Basic introductions from China	1 Hibuna – Common Goldfish introduced *c.* 1502 and/or 1620 2 Wakin – Japanese Common Goldfish 3 Ranchu – Lionhead	4 Ryukin – Fringetail *c.* 1800 5 Demekin – Telescope Meiji era post-1867 ? 1895
Japanese developments from Chinese prototypes	6 Jikin 7 Tosakin	8 Tetsu-wonaga 9 Hanafusa
Group arising from crosses	10 Watonai (Wakin × Ryukin) 11 Kinranshi (Ranchu × Ryukin, Ranchu × Wakin – 1902 Akiyama) 12 Shubunkin 1900 Akiyama	13 Shukin 14 Calico 15 Tetsugyo or Comet
Origin obscure	16 Oranda Shishigashira	

(b) Mitsu-wo (trilobed)
(c) Sakura-wo, intermediate between three- and four-lobed, possibly what we now call web-tails
(d) Funa-wo (single tail). Morphologically the latter is in practice indistinguishable from the Hibuna or common goldfish.

Matsui noted a superior quality of colour in the Wakin *vis à vis* the Hibuna.

3 Ranchu

Under this umbrella designation Matsui includes three major groupings of fish whose common characteristics are the short, broad and rounded body, and the absence of the dorsal fin with the other fins being short. The characters which separate the three groups basically are whether the skull is narrow and pointed (Nankin or Eggfish) without any head growth, whether the skull is broad and blunt, lacking head growth (Maruko or Osaka Ranchu), or whether in addition to the broad and blunt skull there is head growth or hood development (Ranchu proper). The nature of the head growth can be used to differentiate three forms of Ranchu, namely:

(a) The 'Shishigashira' (lionhead) in which development of the hood covers almost the whole head divided into three regions, cranial, infraorbital and opercular.
(b) The 'Tokin' (capped or hooded) in which only the cranial portion is developed.
(c) The 'Okame' (swollen cheeked) in which only development of the infraorbital and opercular areas occurs.

At the present time most lionheads are of the first type but the capped form 'goosehead' is to be seen occasionally while the 'Okame' does not at present find very much favour.

4 Ryukin

This is a short, deep-bodied form with long fins. Matsui also gives the names 'Nagasaki' and 'Onaga' signifying long-tailed. In some individuals the tail was noted as being veil-like. Variation in the extent of duplication of the caudal fin has been recorded, four-lobed, trilobed and bilobed (single-tailed) being noted. The latter were often called 'Fukinagashi' (streamer). The strength of the red colour was noted and the observations made that the head is small and pointed but there may be development of some warty head growth, though not so extensive or prominent as in the Ranchu or Oranda.

5 The Demekin

The Demekin or Telescope-eyed goldfish in Matsui's definition includes a range of forms of which in the view of other authorities actually constitute distinct varieties. All of these are short-bodied fish with paired caudal fins, their most distinctive character being the large and protuberant eyes which vary in shape: globular, cylindrical or pear-shaped. The first three of the forms listed are generally similar but differ in colour or scale type:

(a) The Aka-demekin (Red Telescope). This is a typical red metallic goldfish undergoing the normal process of colour change.
(b) The Kuro-demekin (Moor). At the time when normal decoloration occurs the black colour intensifies greatly, varying finally from 'blackish-brown to smoky-black'. Some individuals retain the wild 'iron' colour while others with age may undergo demelanisation and become red.
(c) The Sanshoku-demekin (Calico Telescope). This variety is characterised by possession of transparent scales together with rather few scattered metallic scales. In the progeny of this variety Matsui recorded that it is not true breeding and produces a 1 : 2 : 1 segregation of all transparent scaled fish (¼), mosaic transparents (½) and normal scaled (¼). The all transparent fish are reported to be somewhat difficult to rear and not maintained as a distinct type. In the mosaic transparent fish the decoloration of the metallic scaled fish does not occur but progressively a mottled pattern of 'red, yellow, blue, violet, white, etc. mingled with black spots' develops. Matsui observed the occurrence of individuals 'either colourless or so pale that it is possible to see through the opercula the colour of blood of the branchia, and the eyeballs sometimes being black from lack of the iridocyst layer in the iris'. Matsui noted that these three variants were imported from Kwangtung, China in 1895, after which they have been bred and multiplied in the country.
(d) The Chôtengan (Deme-ranchu). The importation of what we now know as 'Celestials' first occurred in 1902 or 1903 when Dr Mitsukuri imported about 30 from Kwangtung and gave the name 'deme-ranchu' in 1904. The name 'Chotengan' is close to the original Chinese. The fish itself is described by Matsui as long-bodied, without a dorsal and with longish fins, the most remarkable feature being the eyes, which are upwardly directed. In progeny of this type only about 30% develop the definitive characteristics of the eye fully. The lack of dorsal fin showed a higher development coefficient with 60% of progeny showing the feature.

6 The Jikin (Peacock tail)

This variety has the alternative name 'Kujaku' or 'Kujaku-wo' referring to the form of the tail, meaning literally 'peacock tail'. The caudal fin development is the distinctive characteristic of this type and is quite unique. The caudal fins are duplicated and, viewed from the rear, have the form of an X. The twin caudals diverge strongly and lie almost parallel to each other in a near vertical orientation at their base in the caudal peduncle. In the portion of the caudal peduncle between the tail fins some small scales can develop. Apparently Matsui observed that only about 40% of the progeny of this type showed full development of this distinctive tail form. The general body form is elongated, somewhat similar to that of the Wakin. The desired coloration is of deep red fins borne on a white or silvery body. Since the desired pattern develops naturally in only a small proportion of the progeny, art is frequently called upon to rectify perceived shortcomings of nature and artificial colorants combined with minor surgery are sometimes used to 'improve' fish. This variety is indisputably Japanese in origin; it is most striking but difficult to produce and costly in consequence. The Japanese name is used in the West.

7 The Tosa-kin (Tosa Goldfish)

The Tosakin is a uniquely Japanese fish which has arguably the most beautiful tail developed in any goldfish variety. In general body characteristics it is like a Ryukin with the exception of the caudal fins which are joined along their upper margins so it is a form of 'web-tail'. The caudals extend sideways with a characteristic upward curve of the mid-section; the joined upper section tends to be more or less horizontal. This variety has declined in popularity, perhaps because only about 30% of progeny reproduce the form of their parents. It is not an easy type to breed, which perhaps explains why although it is greatly admired it is not seen very often.

8 The Tetsu-wonaga (Iron-Coloured Long-Tail Goldfish)

This type is basically a metallic Ryukin which retains the wild-type (iron) coloration; de-coloration occurs late if at all. This difference seems a relatively trivial point on which to recognise a distinct variety.

9 The Hanafusa

This appears to be a Japanese counterpart of the Chinese Narial Bouquet or Pompon, from which it differs in possessing a dorsal fin. Matsui suggested that it could have developed from an Oranda which failed to develop a hood but in which the narial septa enlarged to produce ball-like growths.

10 The Watonai

This is the first of a group of varieties which were produced in Japan by hybridisation. The Watonai was reputedly produced in 1883 from the cross between the Wakin and Ryukin. It is intermediate between the parents in both body form and development of its finnage. It is very similar in appearance to the Fantail so very well-known in the West.

11 The Kinranshi

This type was produced by Akiyama in 1902 according to Matsubara from a hybrid of a Ranchu with the Ryukin. The fish has an elongated body lacking a dorsal fin and with metallic scales and dappled. Matsui suggested that the probable parentage was Ranchu × Wakin which he found reproduced the Kinranshi body form with plain red or red and white dappled coloration.

12 The Shubunkin

Akiyama produced this in 1900 from crosses involving the Calico-demekin, the Wakin and Hibuna. The fish has the transparent scales of the Calico-demekin, the body form of the Hibuna and a somewhat longer single tail than the latter. Matsui concluded that the use of the Wakin served no useful purpose and reproduced the type himself from crossing the wild goldfish Funa with the Calico-demekin. This combination genetically provides all that is needed to produce the characteristics of the Shubunkin. The wild-type eyes and single tail of the Funa repress the recessive protuberant eyes and twin tail of the Calico-demekin. The transparent scale character would be expressed in half the progeny of the cross; variation in body length in the progeny would enable selection of the Funa body form to be effected. The length of finnage would be expected to be intermediate, as generally in long × short fin crosses simple dominance does not occur.

13 The Shukin

This is yet another of Akiyama's hybrids, between the Ranchu and Oranda. It combines the dorsal-less condition of the Ranchu with the long finnage of the Oranda; the hood is well developed. The original hybrid line became extinct but Matsui recreated the type, reporting that only 3.21% of progeny were of the desired type.

14 The Calico

Matsui has described a group of hybrids between the Calico-Demekin and the Ryukin, Wakin and Oranda-Shishigashira which produce equal proportions of metallic and mosaic transparent scaled progeny. They are called Calico-Ryukin, Calico-Wakin or Calico-Oranda according to the other parent used. The name was given in 1912 by the American Franklin Pachard. Coloration reported was red and white mottled for the most part, the mottled colour of the Calico-Demekin being rare. Subsequently, however, mottled Calico-Ryukins and Calico-Orandas have been produced in quantity and very good they can be too!

15 The Tetsugyo or Comet

The Japanese name means literally 'iron fish', referring to the wild-type coloration. It is a single-tailed fish with a long body and longer fins than the wild goldfish. In its form it is closely similar to the American Comet which does change colour. Similar fish can be produced by crossing Funa (wild goldfish) with Ryukins. The Comet of commerce is considered an American fish and could have been the outcome of similar chance crosses.

16 The Oranda-Shishigashira

This has been likened to a hooded Ryukin, which it generally resembles in body form and finnage. The body as noted by Matsui was shallower than in the Ryukin. The form described by Matsui was of the Tokin or goosehead type, development of which reportedly was slower than in the Ranchu. Colour was simple or dappled. Mitsukuri believed it to have arisen from crossing the Ranchu and Ryukin but Matsui disagreed. He suggested that it could have been developed by selection in Ryukin lines of those which showed a tendency (which he had observed) to develop warty growth on the head. Experimental crosses carried out by Matsui demonstrated very complex segregations in Ranchu × Ryukin crosses, which on balance Matsui considered would be unlikely to produce the Oranda as we know it. Since his day the Oranda has diversified enormously producing many very handsome variants.

A revised treatment of goldfish varieties is presented by Matsui in his 1972 publication (see Table 3.9).

Matsui, Y. (1972) *Goldfish Guide* (first edition)

The Watonai and Calico are omitted from the 1934 listing, there are included listings of American and Chinese varieties not yet assimilated into Japanese goldfish culture.

The publication in 1972 of Matsui's *Goldfish Guide* was something of a revelation and initiated a revolution in goldfish keeping in the West and brought to the attention of the occidental enthusiast the full range of the riches of goldfish variation in the Orient. China at that time was somewhat eclipsed as the source of information and material of novel goldfish varieties. Since the start of the Second Sino-Japanese War in1931 China has experienced a very troubled period. The Civil War and Revolution which concluded in 1949, followed by the Korean War, the 'Great Leap Forward' and the Cultural Revolution have meant that the only window the West has had on the goldfish world of China has been through Hong Kong. Some influence has also filtered through via Japan.

To return to the Japanese situation, the *Goldfish Guide* presented much information which had been already presented to a scientific audience nearly 40 years previously, much of which had never previously been presented to the popular or lay Western audience. It is instructive and informative to make some comparison between the listings of 1934 and 1972. In general they are markedly similar, with two significant omissions, the hybrid Watonai and the Calico group. It would appear that although the Watonai corresponds closely with the Fantail as recognised in the West, it may have been subsumed in the Ryukin category as the 'Japanese Fantail'. The Calico situation now is that Calico variants of all the original metallic varieties have been produced and are sensibly considered together with their opposite (metallic)

Table 3.9 Listing of goldfish varieties in Matsui's *Goldfish Guide* (1972).

1 Hibuna	9 Tetsuonaga
2 Wakin	10 Hanafusa
3 Lionhead	11 Kinranshi
4 Ryukin	12 Shubunkin
5 Demekin	13 Shukin
6 Celestial	14 Tetsugyo
7 Jikin	15 Oranda
8 Tosakin	

Table 3.9a American varieties.

1 Comet	2 Veiltail

Table 3.9b Chinese varieties.

1 Pearlscale	5 Blue Goldfish
2 Pompon	6 Brown Goldfish
3 Bubble-eye	7 Out-turned Operculum
4 Toadhead	

numbers. The one exception is the Shubunkin which is a rather special case. Although listed still, the Kinranshi and Shukin are thought to be extinct but could be resynthesised from their parental varieties if desired.

The most remarkable change between 1934 and the present has been the phenomenal diversification which has come about especially in the Lionhead, Oranda and the Telescope (or Globe-eye) groups. These are now far from being simple straightforward varieties and will be considered in more detail later.

The consideration by Matsui of the American and Chinese forms not yet assimilated into Japanese culture is of interest. The Comet as already noted has definite affinities with the Tetsugyo, but how the Comet is related to the latter is a matter for speculation. The Veiltail is clearly very much a variety in its own right and now distinct from the parental Ryukin. The Shubunkin is without special honour in the land of its birth and is much more highly regarded in the West than in the Orient. The breeding activities of enthusiasts in Britain mean that there are now three distinctive strains of Shubunkin: the original Japanese style (adopted by the Americans and selected) and the two derived British strains, the London and Bristol Shubunkins. Detailed consideration of all three will be deferred until later.

Chinese varieties considered by Matsui are all mentioned by Chen: the Pearlscale, Pompon (Narial Bouquet), Bubble-eye, Out-turned Operculum, Blue and Brown and a variant form, the Toadhead, of the Bubble-eye. The Pearlscale has become a familiar variety as also have the Bubble-eye and Pompon, so much so that they are among the select group of varieties for which the GSGB has prepared standards. As has already been mentioned, in the West generally the Out-turned Operculum is considered a monstrosity and while it is not unknown, it is certainly not favoured. Chen mentioned that in Bubble-eyes the development of the bubbles may cause a vertical displacement of the eyeball so that the fish resembles the Celestial; however, the structure of the eyes in the two types is not the same. The Toadhead is uncommon and combines a modest-sized infraorbital bubble with an upturned eye. The Blue and Brown metallics are two interesting variants. In addition to the rather light steely blue commonly seen, there has been produced a much stronger blue metallic coloration which tends to shade into black. Brown metallic fish have also become familiar, most notably the Chocolate Oranda. The quality of the colour shows some variation, the shade with a hint of red in it is certainly striking, verging perhaps on the gingery. Further consideration will be given to Chinese varieties when we consider recent literature from China.

Hervey & Hems (1948) *The Goldfish*

It could be said in fairness that between the wars aquarists in the West were largely unaware of the full range of variation that remained untapped in the Orient. Quite soon after the end of the war, in 1948 to be exact, a new book was published in England by G. F. Hervey and J. Hems entitled *The Goldfish*, which has become a classic. It represented a considerable advance on earlier popular literature in presenting accounts of the anatomy, history and varieties of the goldfish. In comparison with earlier work a much more penetrating perceptive was established through the authors' searching and scholarly approach. After the lapse of more than half a century it is still a book that the serious goldfish fancier and breeder should read. Interestingly the authors gave a concise but comprehensive account of the varieties covered by

Chen in the 1920s. They were clearly familiar with not only Chen's work but the early work of Smith in the United States on Japanese goldfish and that of Matsubara and other Japanese authorities. However, neither Wolf's nor Matsui's work is cited and it is possible they were not familiar with these two important authors. This notwithstanding, their contribution to our knowledge and understanding of the goldfish has been enormous.

The listing they give of goldfish varieties is unusually comprehensive for the period, pre-dating as it does the time when fanciers could with relative ease visit Japan, Hong Kong and latterly China itself, by a considerable margin. After a lapse of more than 20 years information on the broader range of goldfish material came into the public domain. Hervey and Hems also review critically previous published accounts of goldfish varieties. They point out the error (already noted) of the misuse of the Japanese name 'Wakin' to denote the 'Hibuna' or single-tailed common goldfish, correctly identifying it as a cultivated form with duplicated caudal and anal fins. They also point out that some names given to varieties can be very ephemeral indeed: in the year of the coronation of King George VI the Shubunkin was promoted as the 'Coronation Fish'. Yellow Common Goldfish have been dubbed 'Canaries' and names of other birds have been applied to new goldfish strains. Hervey and Hems also mention 'Orioles' and more recently 'Blue Jays' have made their appearance. The naming of white goldfish 'Pearls' also mentioned seems to be a cunning commercial ploy to give a certain quality cachet to fish which are little favoured and actually of low commercial value. In the context of Shisan Chen's conclusion that from the Common Goldfish progenitor about 126 breeds of fancy goldfish have been developed, they make the shrewd observation that this number is academic since major varieties may themselves include a number of distinct variants. The corollary of this is to adopt a hierarchical system based arbitrarily perhaps on distinctive characteristics such as the telescope-eye to define a group of varieties which may differ in coloration, length of finnage and body conformation. A similar approach could be adopted for the Oranda and Ranchu, to name but two others.

These two authors have deliberately not produced diagnoses or descriptions of varieties as such. On the principle that one picture is worth a thousand words they reproduced a series of drawings by the late A. Fraser-Brunner of the Natural History Museum to give appropriate indications of what each variety looks like and its principal characteristics. By this means they have avoided paraphrasing the words of their predecessors and given their own most interesting and perceptive comments. It is interesting to list the varieties they actually mention (see Table 3.10) for purposes of comparison with the coverage of earlier authors. Another

Table 3.10 Listing of Goldfish Varieties in Hervey & Hems (1948).

1 Common Goldfish	12 Eggfish
2 Comet	13 Watonai
3 Shubunkin	14 Kinranshi
4 Fantail (Fig. 3.20)	15 Tumbler
5 Veiltail	16 Meteor (Fig. 3.21)
6 Telescope	17 Bubble-eye
7 Moor	18 Pearlscale
8 Celestial	19 Out-folded Operculum
9 Lionhead	20 Sleeper
10 Oranda	21 Blue and Brown Goldfish
11 Pompon (Hanafusa)	

significant point to make is that according to the authors' preface, the actual writing of the book began in 1942 with final publication in 1948. They certainly made excellent use of the war years in preparing the text.

The first ten listed are considered to be the best known in Britain at the time; the remainder had been seen only occasionally in Europe in the flesh or depicted on various works of art. Some of the authors' comments are worth noting. They relate the Veiltail to the Ryukin and mention alternative English names for the latter: Ribbontail, Gauzetail, Lacetail and Class-tail, only the first of these gained very much currency and was at times called the Swallow-tailed Veiltail. It is now customary to make a distinction between Veiltail and Ryukin especially in the light of postwar development of the latter. The range of variation recorded for the Lionhead and Oranda is not very great; the hood of the Lionhead covered the main surfaces of the head while that of the Oranda was of the goosehead type.

Colour variation recorded was wide. The remarks on blue coloration are of particular interest: 'the blue should not be a slate-blue which, in fact, is very common, but a bright forget-me-not blue (known to the Chinese as Kingfisher blue) which is very rare'. Over the past 50 years fish showing good quality blue coloration have become much more common but fish with vivid, vibrant blue are still as highly prized as ever. The Americans are so taken with the blue colour of fish that they often deliberately select against red, which rather goes against the Japanese name for the Shubunkin, for example, which translated means Vermilion Variegated Goldfish. The recent 'Cambridge Blue Shubunkins' seem to be something of a contradiction in terms! In calico fish, patterns of colours develop which could be considered as mottled, spotted or dappled according to size and distribution of coloured areas. The range of colours noted was wide with blue, violet, red, brown and yellow with a scattering of black spots. Colour variation in metallic ('scaled') fish varies; the deep ox-blood of some Comets was noted. This is an interesting colour which is recorded at times only to be lost but which reappears and disappears again. It is tempting to think that intense reds may owe as much to environment and feeding as they do to genetics. Red and white (the white often pinkish), yellow and black (in the Moor) are recorded. There are also good summary accounts of the development of finnages, hoods and eyes.

The account of the lesser known types of goldfish is something of a landmark and includes the latter half of the list. Three of them achieved a modest level of popularity after World War II, namely the Pompon, the Bubble-eye and the Pearlscale. The Blue and Brown Goldfish listed were considered as virtually Blue and Brown forms of the Moor. There are two distinct types of Blue, one is in reality a grey; it is blue in the sense that a Russian Blue cat is blue. The other is a deep, quite lustrous, navy blue which has been seen more often in colour photographs than in the flesh. There is somewhat more subtle variation in the brown colour which may be either a rich gingery brown or a rather nondescript run-of-the-mill shade. Grey and Brown metallic fish these days are most likely to be Orandas.

The Eggfish is another variety which has been seen more commonly in photographs than in the flesh. It is an interesting variety in the context of the evolution of dorsal-less varieties; it could be considered as the basic type of the group, lacking as it does the hood of the Ranchu and the latter's broad, blunt skull; the up-turned, protuberant eyes of the Celestial and the narial bouquets of the Pompon. The similarity between the Watonai and the Fantail (Fig. 3.20) was not remarked; the Fantail is effectively a resynthesis of the evolutionary intermediate stage between Wakin and Ryukin. Hervey and Hems possibly picked up the reference to both

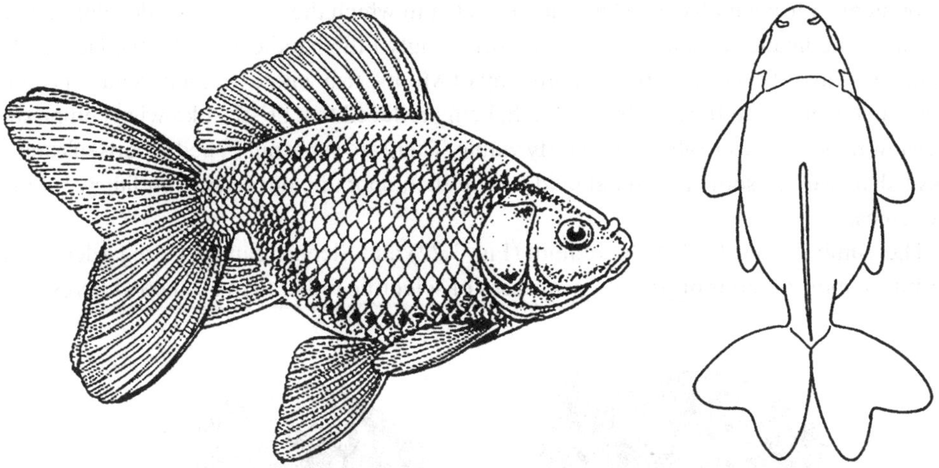

Fig. 3.20 Fantail.

the Watonai and Kinranshi in Yatsu's 1937 report. As noted earlier, the latter was a hybrid which achieved very little popularity and is known only from illustrations in Japanese publications. Interestingly the Pompon illustrated is the Japanese Hanafusa which possesses a dorsal fin, rather than the dorsal-less Chinese form which became established in Britain and for which the GSGB produced a standard, as it also did for the Bubble-eye and Pearlscale.

One of the most intriguing illustrations is that of the Meteor (Fig. 3.21), a deep-bodied metallic fish in which the caudal peduncle and caudal fin have not developed but with the enlarged anal fin in approximately the position of the caudal. In my own experience I have

Fig. 3.21 Meteor. Fish such as this, lacking the caudal peduncle and caudal fin, may swim surprisingly well, the anal in this case being pressed into service as a jury caudal (Hervey & Hems 1948).

encountered a fish which carried no caudal fin but in which there was some development of the caudal peduncle. This individual was surprisingly little handicapped by the lack of this member. I have also come across, in progeny of Moors, fish with extremely reduced caudal fins which were quite functionless as such, being reduced to a bristle-like wisp comprising very few rays. They were also very poorly viable. There are some published reports of caudal-less fish as well as some anecdotal evidence of their sporadic occurrence in fancy goldfish progenies.

The Tumbler, Out-folded Operculum (Fig. 3.22) and Sleeper can all be regarded as deformities, monstrosities or pathological cases of scientific rather than aesthetic interest. The

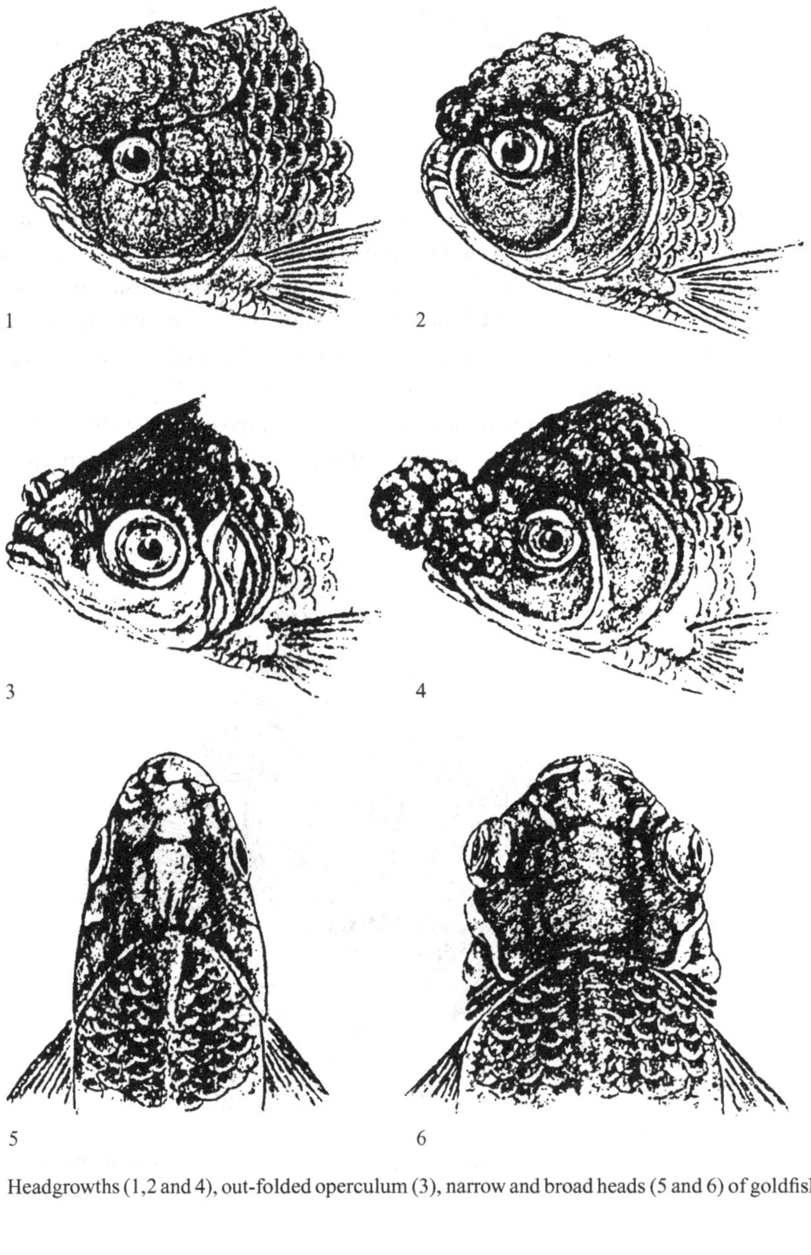

Fig. 3.22 Headgrowths (1,2 and 4), out-folded operculum (3), narrow and broad heads (5 and 6) of goldfish (Chen 1925).

Tumbler is a fish which has a concave back produced by a strong reverse curvature of the vertebral column. It is quite incapable of normal movement, hence its name. Probably this type was produced in progeny of a strain carrying this deleterious mutant, and it is unlikely that such a form would be capable of breeding. The Out-folded Operculum is a mutant form in which development of the operculum is abnormal. If it develops fully then it may curl strongly outwards exposing the gill filaments; in some individuals the development of the operculum may be incomplete, again resulting in exposure of the gill filaments. The Sleeper, which swims on its back, is generally considered now to be a fish suffering from derangement of swim bladder function. Affected individuals are unable to control their buoyancy by exerting pressure on the swim bladder, thus controlling their specific gravity. There is also reference in older works to goldfish 'varieties' which were clearly suffering from dropsy if the characteristic raised scales mentioned in the description are anything to go by.

The situation described by Hervey and Hems was at a critical time when goldfish stocks in Europe were at a very low ebb indeed and some time had yet to elapse before the economics of Europe and the Far East had recovered sufficiently for the trade in goldfish to revive. However, in spite of this precarious situation, the end of the war in August 1945 was rapidly followed by revival of aquarist societies over the whole country and in May 1948 by the establishment of the GSGB.

The coverage of varieties in this book anticipates what was to become commonplace in the 1960s and 1970s when the commercial trade in goldfish was re-established.

Frank Orme (1979) *Fancy Goldfish Culture*

Frank Orme, some 20 years on, developed Hervey and Hems' approach and listed two dozen varieties. It is quite apparent that in his time the range of varieties and variants within the varieties had increased enormously. It is also clear that he had a special interest in the Lionhead group of variants. Table 3.11 indicates an interesting range of material.

Consideration of the succession of variety lists we have looked at so far raises the question of what really counts as a variety. The point is illustrated by the Shubunkin. The London Shubunkin is considered as a transparent scaled (nacreous) form of the Common Goldfish;

Table 3.11 Listing of Goldfish in F. Orme (1979).

1 Common Goldfish	14 Bubble-eye
2 Comet	15 Toadhead
3 London Shubunkin	16 Pompon
4 Bristol Shubunkin	17 Pearlscale
5 Wakin	18 Oranda
6 Jikin	19 Redcap Oranda
7 Fantail	20 Calico Oranda – Azumanishiki
8 Ryukin	21 Lionhead
9 Veiltail	22 Redcap Lionhead
10 Tosakin	23 Chinese Lionheads
11 Demekin	24 Phoenix
12 Moor	25 Meteor
13 Celestial	

the Bristol Shubunkin is considered to have arisen from crosses between British and American Shubunkins which had been selected for long fins. In fact the original Shubunkin as depicted nearly a century ago had longer fins than the Common Goldfish, not as long as the Comet but rather similar in shape. The size and shape of the caudal fin is the hallmark of the Bristol Shubunkin. Its large rounded lobes are unique to the Bristol. Interestingly this characteristic has been established since the end of World War II at which time the tail fins were full but bifurcated as represented in the Federation of British Aquatic Societies (FBAS) standards of 1947, 1954 and 1988. In the GSGB there was a notable change between 1972 and 1986 editions when the new Standard Bristol tail was accepted. It would be sensible to recognise three distinct types of Shubunkin, similar in body form but differing in size and shape of the fins.

Another issue on which Orme takes a rather different stand is the Veiltail, in which he includes not only the metallic and calico forms but also those with telescope eyes. These arguably should be grouped with the Demekins, at least as far as the show exhibitor is concerned – they would be placed in a different class such as the Broadtail Moor, if they were black. With regard to the Orandas, there is a treatment like that of the Shubunkin in that three forms are recognised: metallic, calico (Azumanishiki) and Redcap. The latter has a goosehead type of hood, in which only the cranial area produces hood growth.

The situation in the Lionhead is even more complicated. There are three heads under which Lionheads are considered, the Redcap with the characteristic Tancho red head ideally confined to the goosehead hood, the Lionhead group proper and Chinese Lionheads. The Lionhead group comprises five sub-groups: Ranchu, Osaka Ranchu, Nankin Ranchu, Lionhead and Edo Nishiki. The curious feature about this assemblage is that only the Ranchu (Japanese) and Lionhead (Chinese) groups develop substantial hoods. The Edo Nishiki is the result of an attempt to produce a calico Lionhead but hood growth is generally rather poor. The Osaka Ranchu had a very characteristic globular body, a web-tail and broad head; at one time it was quite popular but is rarely seen if ever at the present time. No hood develops in this type nor in the Nankin which is also known as the Eggfish; it has a narrow pointed head unlike the true Lionheads, and the colour is frequently silver with red fins and mouth.

Orme makes the distinction between Ranchu and Western Lionheads on the basis of the different body forms. The Ranchu is characterised by very strongly curved dorsal and ventral profiles, the caudal peduncle has a slight downward curve in spite of which the caudals are carried in a more or less horizontal plane. There may be a slight asymmetry between dorsal and ventral lobes of the caudals with the length of the dorsal lobes being greater. In contrast, the Western style Lionhead has much less extreme curvature of dorsal and ventral contours than the Ranchu but notwithstanding this a relatively short and chunky body. As in the Ranchu, the fins are relatively small, the caudals have symmetrical dorsal and ventral lobes and are held horizontally in the same plane as the body axis. Chinese Lionheads, according to Orme, have well-developed hoods but are longer in body and fins than the Ranchu and the Western style Lionhead (generally they are closer to the latter than the former).

The Phoenix is an interesting fish which made a strong impression in the 1970s. It is a dorsal-less calico fish with long fins and a rather elongated body. It can be a most elegant fish.

Watanabe (1988) *Handbook of Goldfish*

Watanabe's useful book provides a valuable update of the Japanese perspective on goldfish. The situation has changed greatly with the rise of Koi culture and a very substantial transfer of interest to the development of novel forms of this admittedly interesting and fascinating fish. At the present time the peak of goldfish interest is rather firmly based on the Ranchu which has attracted the interest of specialist societies ('Kai') devoted to its cult and development. While some attempts have been made to develop new goldfish varieties in Japan, these have not been as successful worldwide as those from China although the Hamanishiki is a possible exception. The Edo-jikin which was featured in aquarium publications and which on the face of it was a very attractive fish has virtually disappeared. The purist could argue that it was not in fact a Jikin at all but a Wakin. A more accurate description might have been 'Calico Wakin'. It could almost have been described as a twin-tailed London Shubunkin. This event shows that in the long run success or failure in launching new varieties depends on public taste, which cannot be predicted.

Watanabe lists some 21 varieties (Table 3.12).

The majority of fish on the list have been considered in detail by Matsui. However, there are some which were not described earlier which merit some comment. The 'Blue Marked Goldfish' (Seibun) was imported from China after World War II and has a very striking blue-black colour, often with paler white sides and underparts, which could otherwise be considered as related to the Ryukin group, having normal eyes, but without extreme development of length of finnage or depth of body. The 'Brown Goldfish' has become familiar to us as the Chocolate Oranda. The Japanese name 'Chakin' likens the colour to that of brewed black tea. It also came to Japan from China after the end of World War II and became popular in the West. The development of the hood is variable; those in which it is absent could be considered as Brown Ryukins although without the extreme development of Ryukin features. The Calico is a Ryukin with transparent scales and typical calico coloration, markings and the standard Ryukin conformation.

The Dutch Lionhead is the typical Oranda with metallic scales and well-developed hood. The Dutch Calico or Azuma Nishiki is the Calico Oranda produced by crossing the Calico Telescope and Oranda; some extraordinarily fine examples have been produced by the breeder Azuma. Another very interesting Chinese import was the Redhead (Honto-yui or Tancho) which has become extremely popular in the West where we have seen some extraordinarily

Table 3.12 Listing of Varieties in Watanabe (1988).

1 Comet	12 Dutch Calico
2 Shubunkin	13 Brown Goldfish
3 Wakin	14 Red Head
4 Jikin	15 Buffalo head/Ranchu
5 Ryukin	16 Nanking
6 Blue Marked Goldfish (Seibun)	17 Edo Calico
7 Tosakin	18 Pearlscale
8 Telescope Eye	19 Hama Calico
9 Calico	20 Narial Bouquet
10 Celestial	21 Bubble-eye
11 Dutch Lionhead	

beautiful examples. All in all, one of the most remarkable features of postwar goldfish variety development has been the proliferation of new, beautiful and interesting variants with what has become virtually a varietal family. In addition to production and dissemination of these Chinese types, the Japanese in the Azuma Nishiki have added yet another jewel to the Oranda crown.

The Edo-Calico or Edo Nishiki is in essence a Calico Lionhead; although produced by the cross between the Lionhead and Calico Oranda, the hood development is unimpressive. The genetic contributions of the two parents to hood development seems not to have produced a positive and additive effect on this but rather to have been mutually neutralising. This could be construed as support for Matsui's contention that hood development in the Oranda and Ranchu lines occurred independently. The Hama Calico apparently arose from a goosehead pearlscale line obtained from Hong Kong in 1967 from which was selected in Japan the current 'Hamanishiki' in which the head growth takes the form of a paired bubble. Apparently a similar type was developed in China but most of those seen in the West are of Japanese provenance.

The role of Hong Kong in postwar development of the goldfish can hardly be overestimated. For the West it has been a window to China for several decades through which highly significant materials have passed from China first to Japan and then to the West. The situation in recent decades has changed with resumption of more direct contacts between China and the West which we can now consider.

Pénzes & Tölg (1986) *Goldfish and Ornamental Carp*

Bethen Pénzes and István Tölg are two Hungarians who published their work *Goldfish and Ornamental Carp* in 1983 in German (*Goldfische und Zierkarpfen*) and had an English translation published in 1986. It is very interesting to have contributions from a wide geographical range, otherwise a rather introverted perception can develop as seemed to happen between the wars. The vastly increased mobility of the population at large prevents this happening at the present time.

These authors consider some 22 varieties or groups of varieties, a listing of which is given in Table 3.13.

Table 3.13 Listing of Goldfish Varieties by Pénzes & Tölg (1986).

1	Wild type Orange Goldfish (Common Goldfish)	12	Celestial, Heavenward Star gazer, Sky gazer, Deme Ranchu
2	Comet – Tetsugyo	13	Bubble-eye
3	Shubunkin	14	Eggfish – Maruko
4	Common Domestic Goldfish (Japanese Wakin)	15	White Eggfish – Nanking
5	Peacock Tail – Jikin	16	Osaka Ranchu
6	Arrowtail – Watonai	17	Lionhead, Bramblehead, Ranchu
7	Veiltail, Ribbontail, Fantail, Fringetail, Twintail, Ryukin	18	Oranda
8	Goldfish of Tosa, Tosakin	19	Calico Oranda-Azuma Nishiki
9	Red Telescope Eye, Globe-eye, Demekin	20	Red head (Oranda)
10	Black Moor, Black Telescope Eye, Kuro-demekin	21	Pearlscale – Chunshuyui
11	Calico Telescope Eye – Dragon Eye	22	Pompon (Hana fusa)

One of the most interesting features of this listing is the grouping together of the Veiltail, Fantail and Ryukin. Curiously the Watonai is considered a separate entity, even though in its conformation it is virtually a Fantail. The Watonai which was produced by crossing Wakin × Ryukin is an intermediate between the two parents. The Ryukin is thought to have been derived from the Wakin by long sustained selection, which could have produced the Fantail at an intermediate stage. The Watonai essentially puts the clock back to this intermediate phase, effectively recreating the Fantail. The relationship between Ryukin and Fringetail is quite well documented over the past century. The divergence arising from different selection pressures has produced two very distinct forms which justifiably can be regarded as different varieties.

Relationships between Maruko, Nankin, Osaka Ranchu and Lionhead are not straightforward but there is a sensible consensus that different lines selected from the Maruko produced the Nankin (hoodless like the Maruko) and the Osaka Ranchu (with little or no hood development). It is interesting to note that the Lionheads depicted and discussed are of the Chinese rather than the Japanese Ranchu style. The book was written and published before the fall of the Iron Curtain and as a consequence it is highly likely that difficulty was experienced in obtaining Japanese goldfish of the Ryukin and Ranchu varieties. The authors were able to visit China at a time (1965) when Westerners would have found it very difficult to do so, just before the Cultural Revolution which began in 1966.

Recent literature on Chinese goldfish

There is no doubt that the most accessible goldfish literature is that which has been published in English. It is fortunate that we have some published English translations of works originally published in Chinese, Japanese and German. Fortunately for those whose mother tongue is English, some of the landmark publications in the literature by the eminent oriental scholars of the goldfish Chen and Matsui were originally in English. Unfortunately there is an enormous bulk of literature in both Chinese and Japanese on the goldfish which is all but inaccessible to the Westerner. There are also works published in German, notably those by Teichfischer, which are not available in translation. In the latter case, the situation is easier as it is often possible to get some assistance in making translations or even attempting to do so oneself. It is a very different kettle of fish in the case of oriental language translations; the translator needs to be not only perfectly bilingual in English and Chinese or Japanese but also very knowledgeable on the subject of goldfish. There are those who are well qualified to produce such translations but who lack the time and the motivation to do so. We must therefore make the very best use of what is available and hope in the fullness of time that more translations of interesting observations and experimentation may become available in the future.

It is fitting that the final section of this review should consider work from the Orient (although it has not been sufficiently widely distributed in English), and from Europe, which has not been translated from the German. In this way it should be possible to appreciate how we have arrived at our current state of knowledge.

Man Shek-hay (1982, 1993) *Goldfish in Hong Kong*

Man Shek-hay produced his first edition of *Goldfish in Hong Kong* in 1982 and the second revised edition in 1993 (see Table 3.14). The latter is a benchmark publication with a remarkable collection of fine colour photographs. It was the first postwar publication to present in the English language a picture of the goldfish situation in Hong Kong and China. It is unfortunate that this publication is difficult to obtain outside Hong Kong as it is a very informative book indeed. The relative isolation of China from the West since the early 1930s has meant that not only have we in the West been denied access to much Chinese material but we are unfamiliar with the naming of many varieties, especially new ones. Chinese names can often be long because they attempt to describe appearance; in English and Japanese the names are shorter, often only of a few syllables. Perhaps this explains why, although the Japanese names are current for several varieties in the West, no Chinese ones are in use, even though far more material originated in China. In addition, a different descriptive vocabulary has been used; in English language publications produced in China and Europe this has tended to be literal translations of the Chinese, which do not necessarily mean much to the occidental.

Although Man Shek-hay mentions the curled operculum or out-folded operculum and illustrates the condition, none of the varieties described in the text show this feature. This is perhaps a reflection of the status of Hong Kong as a major exporter of goldfish: the tastes of importing countries determine to a very large extent levels of production of the various varieties. Out-folded opercula are not in favour outside China.

Table 3.14 Listing of Goldfish Varieties by Man Shek-hay (1993).

A Single Tails	
1 Carp type Goldfish (Short fins) Common Goldfish	2 Carp type Goldfish (Long fins) Comet Goldfish and Shubunkins
B Twin Tails	
3 Wenyu – Chinese Fringetails/Fantails	8 Japanese Fringetails
4 Wakin	9 Oranda (High head)
5 Jikin	10 Pompon (Velvety ball)
6 Tosakin	11 Pearlscale
7 Ryukin	12 Pearlscale Oranda (Hamanishiki)
C Telescopes	
13 Dragon-eye	16 Dragon-eye Orandas
14 Magpies	17 Dragon-eye Pearlscale
15 Dragon-eye Pompons	18 Butterfly tails
D Dorsal-less	
19 Eggfish	25 Pompon Lionhead
20 Pompon Eggfish	26 Telescope Lionhead
21 Pompon Dragon-eye Eggfish	27 Ranchu
22 Egg-Phoenix	28 Bubble-eye
23 Redcap Egg Phoenix (Tancho)	29 Bubble-eye with dorsal
24 Lionhead	

Li Zhen (1988) *Chinese Goldfish*

In 1988 there was published in Beijing the first English language book on the goldfish from the mainland. This followed a similar approach to that of Man Shey-hay in 1982, but presumably the author was able to draw on the goldfish resources of the whole of China not just the south of the country, to which people in Hong Kong had the easiest access. The range of variation covered is actually very similar to that provided by Man Shek-hay (see Tables 3.15, 3.16).

In the preparation of the genealogical table presented the authors make an important distinction between mutation and recombination of attributes which enables a fairly clear presentation of the complex variation pattern to be set out.

Perhaps the most interesting feature this listing reflects is that about one-quarter of the types listed feature out-folded opercula or reversed gills. Clearly this selection reflects the taste and preferences of the Chinese who do not have the negative attitude to this feature which is fairly general outside China. The local market obviously makes it worthwhile to produce this type in quantity.

Table 3.15 Listing of Goldfish Varieties by Li Zhen (1988) (basic types).

1 Crucian carp – wild goldfish	11 Egg Fish
2 Golden Crucian carp – Common goldfish	12 Pompon
3 Wakin – twin-tail goldfish	13 Reversed gill Eggfish
4 'Carp-like fish' (Fringetail)	14 Lionhead
5 Goosehead (Oranda)	15 Tigerhead (a Lionhead variant)
6 High head (Oranda)	16 Goosehead Eggfish
7 Dragon-eye (Telescope)	17 Froghead
8 Dragon-back (Dorsal-less Telescope)	18 Bubble-eye
9 Sky-gazing eyes (Celestial)	19 Eggfish with Phoenix tail
10 Pearlscale	

Table 3.16 Listing of Recombinant Variants by Li Zhen (1988).

1 Dragon-eye Oranda	15 Tigerhead Reversed gills
2 Dragon-eye Pearlscale	16 Bubble-eye Goosehead
3 Dragon-eye Pearlscale Reversed gills	17 Bubble-eye Pearlscale
4 Dragon-eye Reversed gills	18 Egg-Phoenix Oranda
5 Dragon-eye Pompon Reversed gills	19 Egg-Phoenix Reversed gills
6 Dragon-eye Pompon	20 Egg-Phoenix Pompon
7 Dragon-eye Pompon Oranda	21 Pompon
8 Oranda Pompon Reversed gills	22 Pompon Reversed gills
9 Oranda Reversed gills	23 Pompon Lionhead
10 Oranda Pompon	24 Dragon Back Pompon
11 Pearlscale Pompon	25 Dragon Back Tiger head
12 Pearlscale Reversed gills	26 Celestial Bubble-eye
13 Pearlscale, Dragon-eye Pompon Reversed gills	27 Celestial Pompon
14 Tigerhead Pearlscale	28 Froghead Pearlscale with Reversed gills

B. Teichfischer (1994) *Goldfische in Aller Welt*

Bernhard Teichfischer has performed a useful service for goldfish enthusiasts in Germany in producing his *Goldfische in Aller Welt*. His studies on goldfish in China, based on his own experience, have resulted in perhaps the most comprehensive account of the variation which has arisen in the country in recent years. As a result of difficulties and restrictions in contact in the last half-century between China and the West, this work helps us enormously to develop a much better informed perspective on the global goldfish situation. The future of the goldfish in terms of progress in the development of new, intriguing and interesting varieties depends on China. Japan, which formerly was a major innovator in the goldfish world, has, since the end of World War II and the development of the cult of Koi, become pre-eminent in this field and the major arbiter of taste as well as an amazingly prolific source of new variants to the detriment of interest in goldfish. In a word, it all now seems to depend on China. Since no translation into English of Teichfischer's work is available, I hope to summarise and comment on his perceptions and perspectives on the goldfish. What he has done in effect is to amplify and collate the information which he has garnered in China and incorporate it into a development of Matsui's goldfish genealogy. This is a very considerable aid to the general understanding of the position for the Westerner in the light of past and recent history.

One major problem has been in the matter of names. It is a matter of record that Japanese names for goldfish varieties have been adopted readily by enthusiasts in the West for several varieties. Good examples are Wakin, Jikin, Shubunkin, Ryukin, Hamanishiki and Oranda. Neither the original Chinese names nor literal translations of these names have proven to be acceptable. This may be because Japanese words are easier to pronounce than Chinese. Chinese names also tend to be long and poetic and sometimes the translations can be unfortunate. However, the ready availability of high-quality colour photographs of fish means that relative unintelligibility of translated names is much less significant now than it was in the first half of the century.

Matsui's presentation of goldfish genealogies has inspired writers on the goldfish to amplify and update. To expand the elegant and straightforward Matsui chart (Table 3.8) to incorporate the range of Chinese variants would be to produce an over-complex and virtually incomprehensible system. What Teichfischer has very sensibly done is to identify lineages in a simple version of the Matsui scheme and identify three lineages in which rather complex patterns of diversification are manifest. Information on lineages is presented in Table 3.17.

There are some points of interest which merit comment. The first of these is the Yamagata goldfish, mentioned by Matsui but which does not figure in any version of his published goldfish genealogies. It is in essence a twin-tailed Comet, a fish which might have been expected to have aroused more interest than it has; it could probably be reconstituted without undue

Table 3.17 Major Lineages Established by Teichfischer (1994).

1	Silbergiebel – silver Crucian carp	6	Pfanschwanz – Jikin
2	Gewöhnlicher Goldfisch – Common goldfish	7	*Fransenschwanz – Fringetail
3	Kometenschweif – Comet goldfish	8	Ryukin
4	Japanischer Goldfisch – Wakin	9	*Teleskop – Fäscherschwanz – Fantail Telescope
5	Yamagata Goldfisch – Yamagata	10	*Eierfisch – Eggfish

* highly diversified lineages

difficulty by crossing the Comet and Wakin and selecting for twin tails and long finnage in the progeny.

It might be argued that the Ryukin could have been considered as part of the Fringetail lineage but equally the Japanese Ryukin in its fullest development could be considered an end-point or dead-end in selection and regarded as a more or less definitive variety like the Jikin, unlikely to evolve further and diversify extensively. The products of known hybridisation such as the Shubunkins are not considered in the initial patterning of the variation but in relation to the parent which contributed their most distinctive characteristic.

It is useful to consider how Teichfischer handles the three lineages (Tables 3.18, 3.19, 3.20). This can be considered at two levels: the first encompasses recognition of distinct divergent lines and the second the variations on these major themes. These variants are commonly forms which have acquired through hybridisation features such as hoods, pompons, out-turned opercula, transparent scales and pearlscales. By appropriate crosses and selection the Chinese have shown that it is possible to produce combinations of almost any characters. Since most of these are on the head, it is a moot point whether aesthetically the effort is worthwhile, although one must acknowledge that to accumulate three such attributes is no mean achievement.

Teichfischer has wrestled with the problem of producing a logical and consistent approach to the presentation of an account of the bewildering array of varieties which has been produced not only in China but also in Japan and the West. To be totally consistent is perhaps an impossibility, especially in view of recent Chinese practice in attempting to produce new variations on established themes noted earlier.

Table 3.18 The Fringetail lineage in Teichfischer (1994).

1 Fringetail	7 Japanese Narial Bouquet – Hanafusa
2 Fantail	8 Iron Fringetail
3 Watonai	9 Meteor
4 Veiltail	10 Tosakin
5 Goosehead Oranda	11 Ironfish
6 Oranda	

Table 3.19 The Telescope-eye lineage in Teichfischer (1994).

1 Telescope Fantail	4 Butterfly Telescope
2 Calico Telescope-eye	5 Telescope Eggfish
3 Telescope Veiltail Moor	

Table 3.20 The Dorsal-less lineage in Teichfischer (1994).

1 Eggfish	8 Eggfish with out-turned operculum
2 Lionhead	9 Bubble-eye
3 Ranchu	10 Bubble-eye with dorsal
4 Goosehead Eggfish	11 Toadeye Eggfish
5 Nankin	12 Pompon Eggfish
6 Osaka Ranchu	13 Goosehead Telescope-eye Eggfish
7 Egg-Phoenix	

It is presumptuous to criticise but a possible alternative way of presenting a scheme for orderly classification of goldfish varieties might be to accept a basic scheme presenting relationships between major varieties and identifying groups which have arisen through crosses and recombination (see *Chinese Goldfish,* Table 3.16) and list these according to major features. Such groups would overlap to some extent but it would be useful to have some kind of organised catalogue of variants. Tables 3.21–3.26 represents an attempt to do so.

Table 3.21 Pearlscale variants in Teichfischer (1994).

1	Telescope Lionhead Pearlscale	7	Pearlscale with out-turned operculum
2	Telescope Out-turned operculum Pearlscale	8	Tigerhead Pearlscale
3	Telescope Pearlscale	9	Bubble-eye Pearlscale
4	Oranda Pearlscale	10	Toadhead Pearlscale
5	Pompon Pearlscale	11	Toadhead Pearlscale with out-turned operculum
6	Pompon Pearlscale with out-turned operculum	12	Pearlscale

Table 3.22 Pompon variants in Teichfischer (1994).

1	Pompon Telescope Fantail	7	Pompon Pearlscale
2	Pompon Telescope Fantail with out-turned operculum	8	Pompon Pearlscale with out-turned operculum
3	Pompon Telescope Eggfish	9	Pompon Fringetail
4	Pompon Celestial	10	Pompon Lionhead
5	Pompon Oranda	11	Pompon Eggfish
6	Pompon Oranda with out-turned operculum	12	Pompon Egg Phoenix

Table 3.23 Out-turned Operculum (OTO) variants in Teichfischer (1994).

1	Pompon Telescope with OTO	7	Pearlscale with OTO
2	Telescope Pearlscale with OTO	8	Oranda with OTO
3	Telescope Fantail with OTO	9	Lionhead with OTO
4	Telescope Butterfly with OTO	10	Eggfish with OTO
5	Pompon Oranda with OTO	11	Pearlscale Toadhead with OTO
6	Pompon Pearlscale with OTO	12	Egg Phoenix with OTO

Table 3.24 Hooded variants – Orandas, etc. – in Teichfischer (1994).

1	Oranda	5	Oranda with OTO
2	Goosehead Oranda	6	Pompon Oranda
3	Broadtail Oranda	7	Pompon Oranda with OTO
4	Calico Oranda	8	Goosehead Telescope-eye

Table 3.25 Dorsal-less Hooded variants – Lionheads, etc. – in Teichfischer (1994).

1	Lionhead	7	Lionhead with OTO
2	Ranchu	8	Tigerhead Pearlscale
3	Shukin	9	Goosehead Bubble-eye
4	Goosehead Egg Phoenix	10	Bubble-eye Celestial Fringetail
5	Goosehead Eggfish	11	Goosehead Telescope-eye Eggfish
6	Pompon Lionhead	12	Kinranshi

Table 3.26 Single-tailed variants (arising from twin-tailed stocks, crosses) in Teichfischer (1994).

1	American (Japanese) Shubunkin	4	Ironfish, Eisenfisch, Tetsugyo
2	London Shubunkin	5	Nymph
3	Bristol Shubunkin		

Teichfischer has attempted successfully to produce a more or less global conspectus of variation within the goldfish species. While the picture which emerges is complex, it is possible that further consideration and discussion might bring about some clarification in our overall perspective. This would be chiefly in the weighting of character differences. I think it is fair to say that the basic types of goldfish have been established and only relatively minor variations are likely to arise in the foreseeable future, but such changes, though relatively slight, can be of enormous significance. Good examples are the relatively minor changes in carriage of the caudals producing the 'Butterfly-tails' and the production of the very striking Magpies and Pandas in which efforts to fix what have generally been regarded as transitory colour combinations have achieved at least a measure of stability.

In what follows I hope to elaborate and discuss further the current variety position in goldfish and produce a broad consensual synthesis in describing current variation in the goldfish.

Chapter 4
Modern Goldfish Varieties

The spectrum of goldfish varieties has changed appreciably in the post-World War II period. We have now virtually the complete range of the major types which have been developed over the centuries in the Far East. Developments in recent times have been in the production of variations on established themes, so much so that in some instances what in pre-war days was a fairly straightforward variety has become virtually a family of sub-varieties. This has happened most notably in the Oranda in which some outstanding new variants have become popular, similarly in the Telescope group the development of Butterflies, Pandas and Magpies has produced some highly acceptable and beautiful types. The changes have also been rung on the Lionhead. A very important recent development has been in what can be regarded as an extension of the Japanese initiative of a century ago in the production of new forms which combine characteristics of two or more basic varieties. It is difficult these days to find among imported fish a Celestial which is also not carrying pompons. Although they are not at all popular in the West, the out-turned operculum has been introduced by crossing and recombination to most Chinese types. Perhaps I am not alone in thinking that at least some of these efforts are misguided. An Oranda or Lionhead is surely carrying an adequate sufficiency of special features on the head without having to support pompons in addition. The most important point that one can make is that the head can only support one or exceptionally two special characteristics and show these off to the best advantage. It may well be a challenge to the breeder to see just how many bells and whistles can be crammed onto the head of a goldfish but many if not most would ask him to spare a thought for the fish. Interestingly, according to the practice of the Goldfish Society of Great Britain (GSGB), fish showing features in addition to the specified characteristics of the variety in question would be ineligible for entry in the relevant show class.

Notwithstanding these reservations, recombination of varietal characteristics has a very positive aspect. This is how calico variants of all the major basic varieties have been produced and there is no good reason why the practice should not be encouraged provided that the characteristics in question can be developed and manifested in a becoming manner and are not detrimental to the fish which bears them.

The Common Goldfish

The Common Goldfish (Fig. 4.1; Table 4.1; Plates 1 and 2) retains the body form and fins of

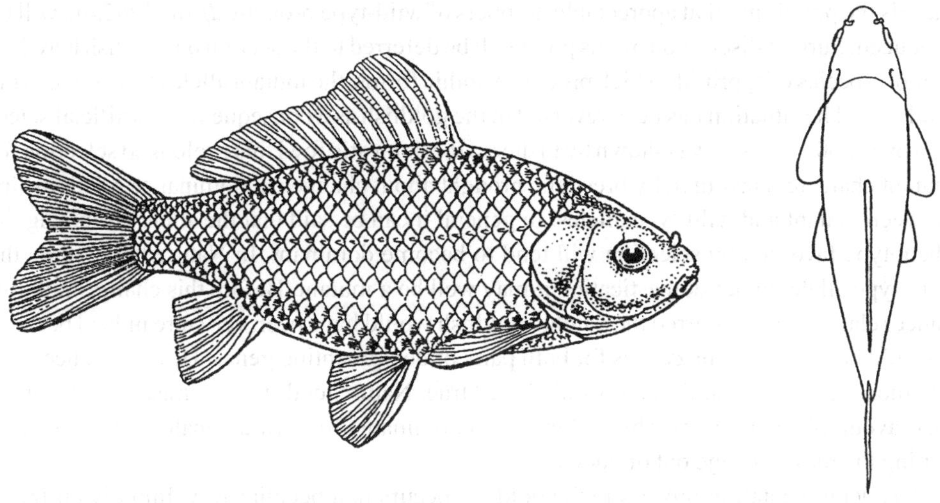

Fig. 4.1 Common Goldfish.

the wild-type fish but differs from it in coloration. Most commonly this is red or orange but yellow fish although much less common are highly regarded; silver or white fish are held in low esteem, however. Occasionally individual fish are seen which are blue, brown or black. In addition to self-coloured fish, variegation is very common. The red and white combination is common and popular; red and black is regarded as transitional and such fish most commonly lose their black coloration to become entirely red. There are indications that it may be possible by patient selection to produce stable combinations of red and black which could be very striking indeed. Blue, brown and black goldfish may occur but are not greatly favoured in a variety favoured primarily as a pond-fish because they would be relatively inconspicuous. The basis for colour differences is primarily presence or absence of one or both of the basic pigments, melanin and xanthine, and secondarily their distribution and the balance between them.

One of the most serious problems that the goldfish breeder faces is that his stock may not breed true. There is a tendency for reversion to the wild type. In the Common Goldfish this is a consequence of the genetic control of the depigmentation trait. Kajishima showed that loss of melanin which results in production of xanthic fish (i.e. red, orange or yellow) is controlled by two genes, *Dp1/dp1* and *Dp2/dp2*. The dominant (*Dp1* and *Dp2*) alleles produced xanthic ('coloured') progeny. Unless either *Dp1* or *Dp2* (or both) are fixed in the parental generation

Table 4.1 Description of the Common Goldfish.

Body:	Elongated, laterally compressed, depth 3/7–3/8 body length.
Fins:	Dorsal and anal fins single, pectoral and pelvics paired. Caudal fin single, spread approximately equal to body depth, lobes approximately 1/4 body length with somewhat rounded ends. Dorsal fin about 3/8 body depth.
Eyes:	Normal.
Contours:	Smooth with approximately equal dorsal and ventral curvature, smooth lateral contours.
Coloration:	Metallic, red, orange, yellow, blue, brown, black, silver, self-colours and variegated. Fins same colour as body or contrasting.

there is the possibility that appreciable numbers of wild-type progeny *dp1dp1dp2dp2* will be produced. Further discussion of this point will be deferred to the section on goldfish heredity. In most species of cyprinids which produce xanthic forms, the mutant allele is recessive to the wild type. The situation has been reversed in the goldfish as a consequence of artificial selection over 1 000 years. It was shown by Fisher in the domestic fowl that prolonged selection for mutant characters favoured by breeders, brought about changes in dominance relationships between mutant and wild-type alleles. In general it can be said that the allele producing the phenotype favoured by selection will tend to become dominant. In Nature this will be the wild-type allele; under domestication it may well be a mutant. Before this change in dominance relationships occurred the xanthic goldfish would have been very rare indeed because it would have to be homozygous for both pairs of depigmenting genes. However, once such a homozygote was established it would breed true. Nature tends to maximise the advantage that favourable genes confer by making them dominant. To draw an analogy it is a way of getting the most mileage out of good genes.

The depigmentation process in the goldfish occurs in a peculiar way. Initially all fry of the Common Goldfish have wild-type coloration. Onset of depigmentation is indicated by a general darkening of colour, there is an intensification of melanin production which is followed by its loss and a consequent lightening of colour from the belly upward. This is brought about by progressive loss of melanophores (containing the black pigment, melanin) which is usually complete resulting in red, orange or yellow fish. This may be followed by a varying degree of loss of the xanthophores (cells containing red, orange or yellow pigments). If there is little or no loss the fish will be self-coloured; partial loss will produce variegation, red and white (or silver), while complete loss will result in a silver or white fish. It is noticeable that fish which lose their melanin first also tend to lose some or all of their xanthic pigments too.

One of the great attractions of the Common Goldfish is the reflectivity of the scales; this is produced by crystalline deposits of guanine in a special class of pigment cells, the iridophores, which constitute a third type of pigment cells. While guanine is not strictly a pigment itself, in the absence of melanophores and xanthophores, it is responsible for the silvery or white appearance seen in variegated or self-coloured individuals. Self-coloured silver or white fish are not especially interesting; a small number in a community of brightly coloured fish does provide a pleasing contrast; However, in combination with highly coloured fins, the combination of silver body plus red fins can be very striking and effective.

As a show fish, the Common Goldfish can be undervalued but to the knowledgeable an outstanding Common Goldfish is something to be treasured, even to the extent of receiving the 'Best in Show' accolade. I had the pleasure of judging the Common Goldfish class at a recent GSGB show; a lemon yellow Common Goldfish entered by Mr Alex Stephenson was placed first in its class and received the 'Best in Show' award. This was as near a perfect example of its kind as one might ever hope to see; in its colour, conformation and condition no fault could be found with it. Contrary to popular belief, perfectly formed Common Goldfish in excellent condition are by no means commonplace. Perfection should be the aim of the breeder whatever variety is being produced, and it is to be hoped will be recognised wherever it is to be seen.

The Comet

The origin of the Comet (Fig. 4.2; Table 4.2) as we know it has been traced by Innes (1947) to the activity of Mr Hugo Mullertt who made selections from the goldfish populations in the ponds of the Fish Commission in Washington during the early 1880s. If one sets this against the record in 1874 of Admiral Ammon's importation of goldfish to the United States it certainly leads one to believe that this could scarcely have been the original introduction but probably was the first reliably recorded importation.

It is interesting to compare the Comet as we know it with the fish described by Matsui, the Tetsugyo or Ironfish, from Japan. This seems to be very similar to the Comet in everything but colour. It is tempting to speculate whether the goldfish used in stocking the Fish Commission included some Tetsugyo and that the Comet was the result of crossing between the latter and the Common Goldfish and recombination in the progeny followed by selection at the hands of Mr Mullertt. Be this as it may, the Comet is a handsome fish indeed and well named. Body form generally is similar to the Common Goldfish except that body depth generally is rather less. In the best specimens the length of the caudal fin should be equal to that of the body. The Comet, because of its bodily conformation and in spite of its long caudal, is very active and fast-moving and is an ideal pond fish. It should never be kept in a small aquarium but always be given plenty of room.

Like the Common Goldfish, the potential range of colour variation is wide but most commonly one sees self-coloured red metallics or variegated red and white metallics. The latter are commonly marketed as Sarassa Comets and, depending on the pattern of the variegation, may be very striking. A particular type of red and white Comet worthy of special mention is the Tancho Comet. This has the large red spot on the head, similar to that on the Japanese crane named Tancho, which is particularly prized by patriotic Japanese because of its similarity to their national flag. The parallel between Sarassa goldfish and Kohaku Koi is quite close; it would be interesting to see whether the elegant patterns and colour quality of Kohaku could be reproduced in Sarassa goldfish by appropriate selection.

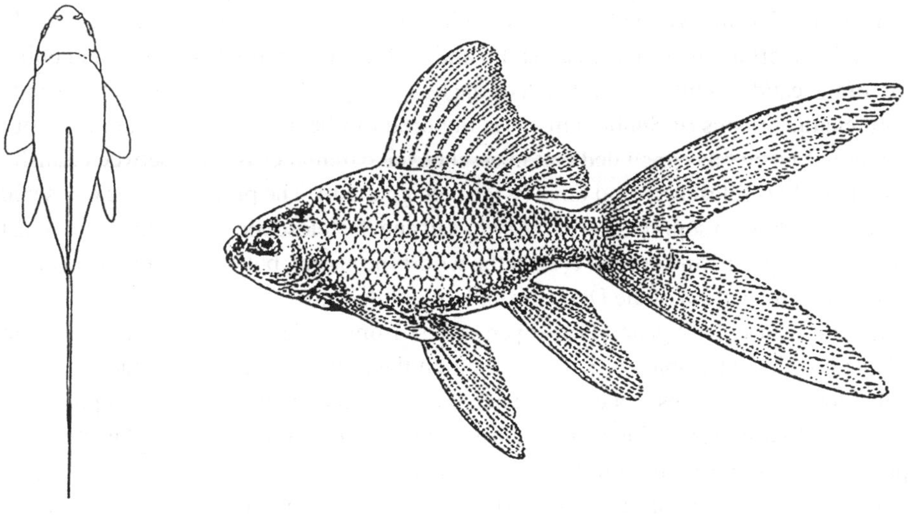

Fig. 4.2 Comet.

Table 4.2 Description of the Comet.

Body:	Elongated, laterally compressed, depth approximately 1/3 body length.
Fins:	Dorsal and anal fins single, pectorals and pelvics paired, caudal fin, long equal in length to that of the body or longer, held well spread. Fins and lobes long, narrow and pointed, dorsal as deep or deeper than body.
Eyes:	Normal.
Contours:	Smooth with approximately equal dorsal and ventral curvature, smooth lateral contours.
Coloration:	Commonly red, orange or red and white variegated (Sarassa or Tancho with white body and red head), other metallic colours possible. Fins same colour as body or contrasting.

It is rather sad that at the present time, although the Comet is a commercial fish, the quality leaves a great deal to be desired. Body depth may be excessive while the caudal fin is commonly no more than half the length of the body. Perhaps some enterprising breeder might consider the possibility of producing premium quality Comets and explore the feasibility of producing also Mirror-scaled Comets (which Merlin Cunliffe has reported from Australia) and even re-synthesising the Japanese Yamagata Goldfish as a twin-tailed version of the Comet.

The Shubunkins

The Shubunkin (Figs 4.3, 4.4, 4.5; Table 4.3) can no longer be considered as a single variety. From a show point of view, in Britain there are separate standards for the two English styles; curiously there is no standard for the original Japanese form. In North America selection has produced a distinctive style of fish (Fig. 4.3) derived from the Japanese original.

The feature which all Shubunkins have in common is that they do not have, for the most part, reflective metallic scales, and they are ideally multi-coloured with strongly mottled or variegated colour patterns. The body form of all variants is similar and comparable to that of the Common Goldfish; excessive body depth is unacceptable. The chief variable between the styles is in the finnage. The original Japanese Shubunkin has longer fins than the standard Common Goldfish and while of generally similar shape to those of the Comet, i.e. with long and sharply tapering lobes on the caudal fin and with the other fins also longer and pointed, length of the caudal is little more than half the body length. In considering the form and the general characteristics of Shubunkins, it is important to bear in mind the parentage of the Shubunkin. Matsui has concluded that in essence the parental cross was between Sanshoku Demekin (Calico Telescope) and Funa (wild-type goldfish). The progeny of the cross which were selected showed suppression of the twin-tail and the telescope eyes of the Demekin, the body depth, single tail and normal eyes of the wild-type parent, but with the expression of the calico colour and pattern of the Demekin and its length of fins.

The American style in general has diverged less from the Japanese than have the London and Bristol styles. The objectives of selection on the part of American breeders have been to select for longer fins, especially the caudal, and against the presence of red pigments in the calico colour pattern. This is a little ironic in view of the fact that the Japanese name emphasises the colour vermilion! In recent years there has been an interesting development arising from the fact that, like all calico fish, Shubunkins do not breed true. They are actually heterozygous for the transparent scale gene, two alleles can occur at a single locus (*T/t*) so

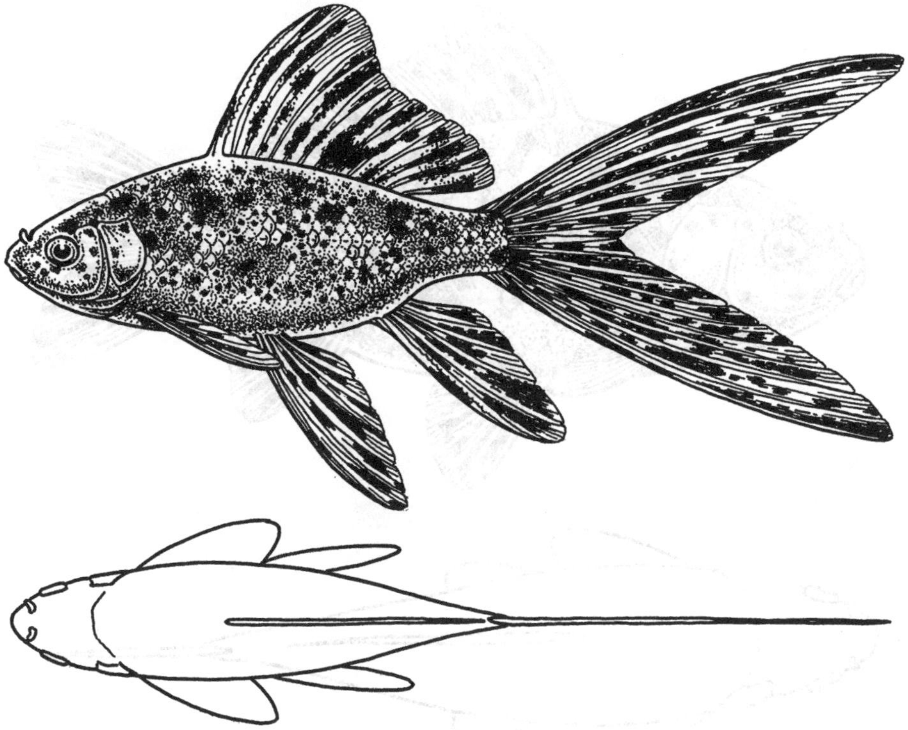

Fig. 4.3 Japanese–American Shubunkin.

when a heterozygote spawning occurs only one half of the progeny will also be heterozygous (and calico in phenotype), one quarter will be homozygous recessive (*tt*) with normal metallic scales while the remaining quarter will be homozygous mutant *TT*. The phenotype of these is termed matt and commonly development of all three types of chromatophore is inhibited. This suppression is not always complete and in some matt individuals some development of both melanphores and xanthophores occurs. Selective breeding of such individuals has resulted in the production of highly coloured, true-breeding (i.e. genotypically matt) individuals. The suppression in part of the effect of the mutant gene in inhibiting the development of two types of pigment cells may be accompanied by partial release of the inhibition of iridophore formation. Thus we can have homozygous individuals which can breed true for the majority of the distinctive characteristics of the Shubunkin. Mr Al Thomma has produced homozygous individuals which closely resemble Shubunkins in all but the development of a liberal sprinkling of dark spots. Commercially the Hunting Creek Fisheries of Maryland have developed highly coloured matt forms, the 'Sky Blue' and the 'Midnight', which are as might be expected blue and black respectively. It would be a very small step in all probability to produce by further breeding and selection a Shubunkin, showing full development of its special characteristics which breed as true as any goldfish variety can.

The two English styles of Shubunkin were developed between the wars. Anecdote suggests that the Shubunkin was introduced and became very popular in the early 1920s. According to Orme (1979) the London Shubunkin (Fig. 4.4) was developed by Messrs L. B. Katterns and A. Derham. The Bristol Shubunkin (Fig. 4.5) was developed by the Bristol Aquarist So-

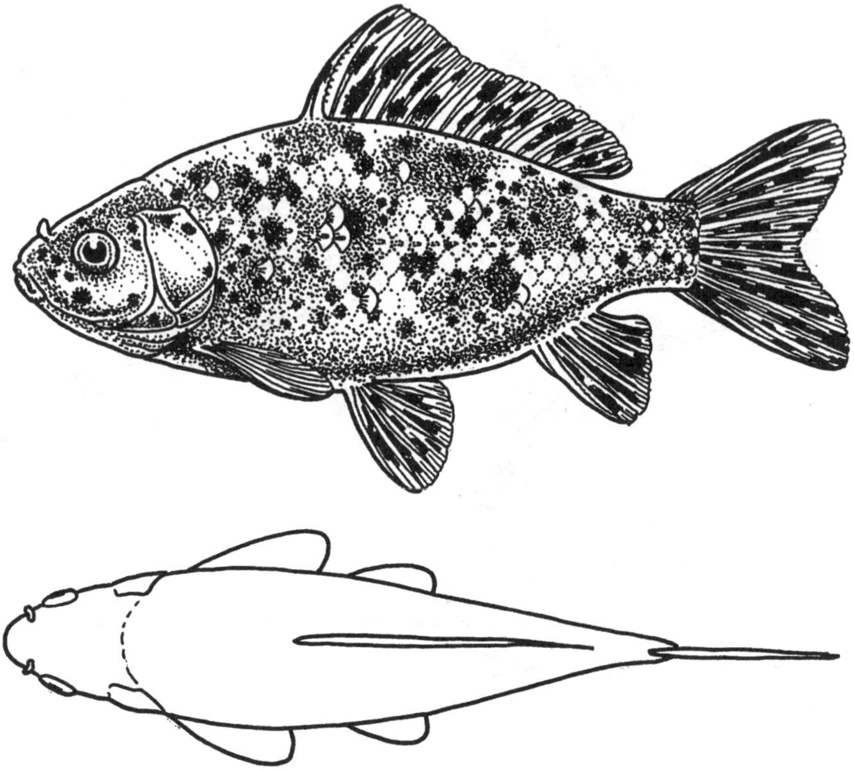

Fig. 4.4 London Shubunkin.

ciety, which published a standard for it in 1934. While the London Shubunkin can be regarded as a calico form of the Common Goldfish, it is not in fact the basic type of Shubunkin as some have assumed. It was probably selected and stabilised after a cross between Japanese Shubunkins and Common Goldfish. Many British fanciers consider it to be but a poor shadow of its Bristol namesake but anyone who has seen the splendidly coloured fish produced by Mrs Pam Whittington, Mr Alan Ratcliffe and the late Mr Bill Leach would agree that the London Shubunkin is no mean fish. Like the Common Goldfish, a good London Shubunkin can carry all before it; Mr Ratcliffe, for example, produced an example which won every class in which it was entered and awarded 'Best in Show' at a recent GSGB Annual Show.

The Bristol Shubunkin at its best is a truly magnificent fish. Like the Japanese Shubunkin it is long-finned but whereas the finnage of the Japanese fish is pointed with narrow caudal lobes, those of the Bristol are rounded with a full tail. Between the wars the tail fin shape was not unlike that of some of the twin-tails, long, full but bifurcated as the pre-war and earlier postwar standards show. In the third edition of the GSGB standards, the old style tail figures, in the fourth edition (1987) the new style is depicted in which the lobes of the caudal each almost describe a circle. They have sometimes been described as like two pennies. A well-grown Bristol Shubunkin showing well-developed finnage and coloration in the peak of condition is truly a sight to behold. The Bristol Shubunkin has in effect been the product of the Society over the greater part of the twentieth century; unfortunately its history and mode of development have not been fully recorded. It seems probable that somewhere along the line

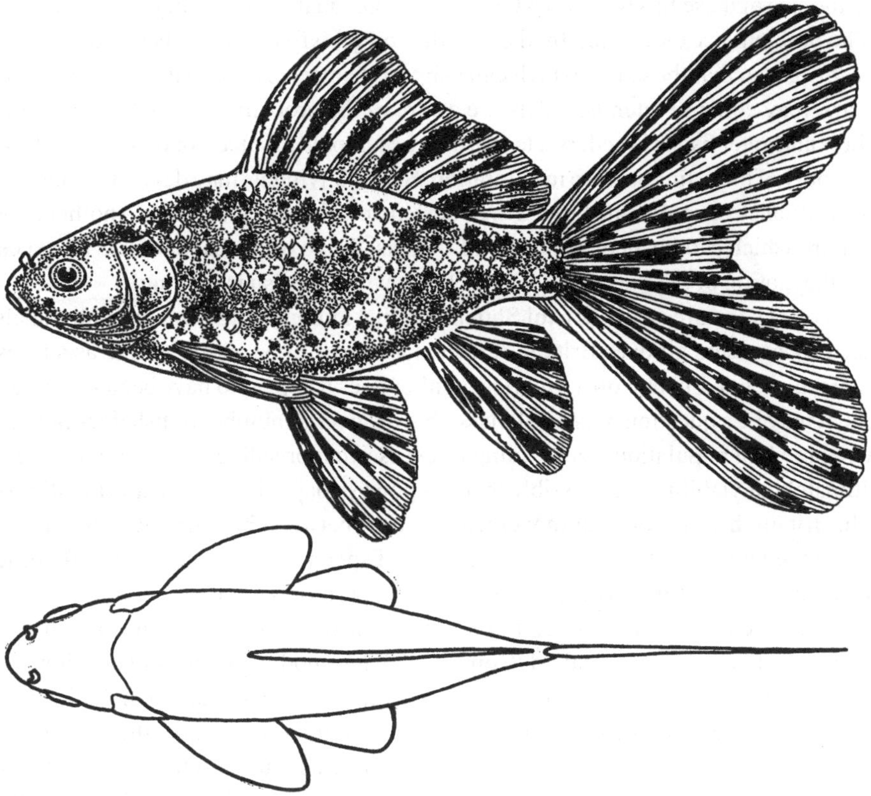

Fig. 4.5 Bristol Shubunkin.

a cross was made into a twin-tailed stock to effect the changes in the finnage. We may never learn just how this splendid transformation was achieved.

The progress which has been achieved in development of the Bristol Shubunkin has also shown that selection for increased length of dorsal and caudal fins can have untoward consequences. It has been found that the transparent scale mutation has the effect of reducing the strength of the spiny rays which support these two fins. This implies that there is an upper

Table 4.3 Description of the Shubunkin.

Body:	Eongated, laterally compressed, depth 3/7–3/8 body length.
Fins:	Dorsal and anal fins single, pectorals and pelvics paired in all, caudal (London) single, spread approximately equal to body depth, lobes approximately 1/4 body length with rounded ends, fins of same size as in the Common Goldfish, caudal (Japanese i.e. original) narrow, strongly bifurcated and pointed, 1/2 to 2/3 body length, somewhat longer in American style fish, other fins long and pointed, dorsal fin height 3/4 body depth, caudal (Bristol) full with rounded lobes 1/2 to 2/3 body length, other fins long but broadly rounded, dorsal fin height 3/4 body depth.
Eyes:	Normal.
Contours:	Smooth dorsal, ventral and lateral contours as in Common Goldfish.
Coloration:	Calico colours and pattern, blue background with patches of red, orange and yellow overlain with even pattern of dark brown or black spots and patches. These features are associated with transparent scales.

limit to the size of these fins which can be supported adequately when they have achieved full size. The problem is exacerbated by the fact that fin growth outpaces that of the rest of the body with age, so that those fins which can support themselves adequately at the age of say two years, may collapse later. Breeders are aware of the problem of over-finnage but those who show and enter fish in breeders' classes may be tempted to select for precocious development not only of these fins but also in coloration. Such individuals may develop collapsed fins and lose brilliance of colour and be past their best in only three or four years. Another undesirable feature which can be a consequence of increased size of the caudal fin is an excessive scissor-like motion of the dorsal and ventral lobes while the fish is moving.

It is a curious feature of the Bristol Shubunkin that although high-quality individuals are produced by breeders all over Britain, it does not appear to be anything like as successful elsewhere, most notably in North America, neither does it appear to have been a commercial success. It may be a question of critical mass, that the overall number of fish does not give an effective breeding population size, leading to excessive inbreeding, loss of vitality, fertility and ultimately of viability. It is possible that it is poorly adapted to American climatic conditions. In Britain there is constant movement of stock between breeders, which may not be feasible in the United States in view of the wider dispersion of breeders and difficulties in making contact. From time to time there is talk and promotion of Calico Comets or Comet Shubunkins. In commercial practice the ordinary Japanese Shubunkin has a comparable length of caudal fin to many so-called 'Comets'. There is nothing to prevent development of the Japanese–American style of Shubunkin into a real Comet Shubunkin or Calico Comet. However, to achieve credibility the length of caudal would have to equal the body in length and be held as it is in the Comet and not drape downwards. Further changes might be rung on the caudal fin. From crosses involving broad-tail twin-tails and Shubunkin it is possible to find progeny which could perhaps be developed into a new style of Shubunkin, a Broadtail, which might well have a similar general conformation to the Veiltail guppy.

Intensity of coloration and the presence of a substantial area of blue colour are features of the best modern Shubunkin of any style. The situation at the end of World War II and in the early postwar period in Britain was that blue Shubunkins were not at all common, and it took a while before they became commonplace. The reason for this in all probability was the presence in the stock of active depigmenting alleles arising perhaps from outcrosses to common goldfish stocks during the war or earlier. This was certainly a feature of London Shubunkin stocks which were most readily available at that time.

The Wakin

The Wakin (Fig. 4.6; Table 4.4; Plate 3) reputedly has been the commonest form of the goldfish in Japan in the past; whether it is so at the present time is a very moot point indeed. Basically it can be considered as a Common Goldfish with twin tails and double anal fins. The body proportions in the young fish are basically the same in the Common Goldfish and the Wakin; however, in mature fish the Wakin gives the impression of having a slightly heavier and deeper body. The fully-fledged Wakin as we know it today was the product of long continued sustained selection for the fullest development of perfectly duplicated caudal and

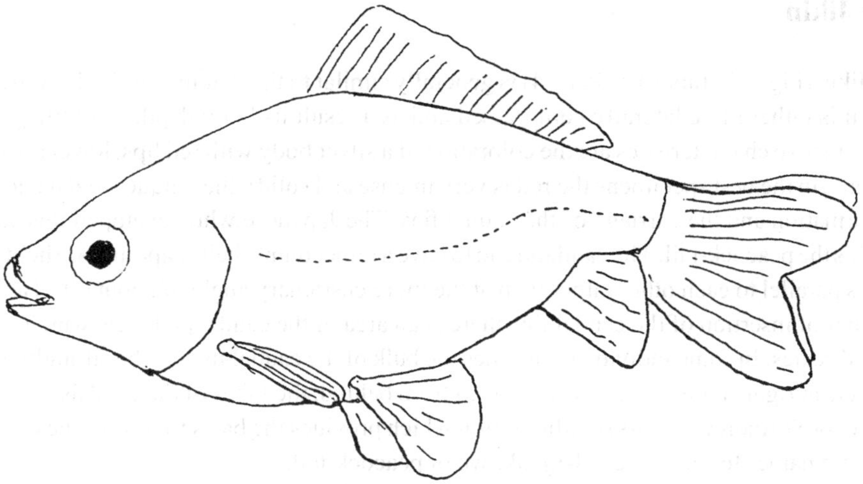

Fig. 4.6 Wakin (J.H. Bundell).

anal fins. This is the conclusion that can be drawn from examination of fish portrayed in eighteenth-century art works when perfectly duplicated caudals were the exception rather than the rule. In any event the Wakin can be considered as the progenitor type of all the exotic fancy goldfish we have today and in no way should it be considered a 'has been' but rather entitled to a place of honour among the goldfish we keep.

The Wakin, according to Japanese writers such as Matsui, has been greatly prized in Japan on account of the very intense and strong development of the red coloration. This seems to be particularly so in the case of variegated red and white fish. This is a feature of Sarassas and other red and white fish; it is almost as though suppression of pigment development in one area is offset by an enhancement or stimulation of its production elsewhere. Typically the Wakin has metallic scales and if desired could be produced, if necessary, by appropriate crosses and selection with any coloration, yellow, brown, black and blue. Since the war a calico long-bodied twin-tail was produced and promoted as the Edo-jikin. This was an extremely attractive fish which failed to achieve any significant measure of popularity. Although called a Jikin, the purists argued that it was nothing of the sort and did not have the very distinctive caudal fins of the Jikin but had caudals which were much closer to those of the Wakin; at best it is an intermediate form.

Table 4.4 Description of the Wakin.

Body:	Elongated, laterally compressed, depth 3/8 to 1/2 body length.
Fins:	Dorsal fin single, anal, pectoral and pelvics paired, caudal fins double, spread approximately equal to body depth, lobes approximately 1/4 body length with somewhat rounded ends, dorsal height 3/4 body depth.
Eyes:	Normal.
Contours:	Smooth with approximately equal dorsal and ventral curvature and smooth lateral contours.
Coloration:	Metallic commonly red, orange and red and white.

The Jikin

The Jikin (Fig. 4.7; Table 4.5; Plate 4) is generally similar to the Wakin in its body conformation; it is rather more laterally compressed and as a result its body depth is a little greater. Its distinctive characteristics are the coloration of a silver body with red lips, lower abdomen and fins (in the best specimens the red is very intense and solid), the unique caudal peduncle conformation and the carriage of the caudal fins. The Japanese who developed this variety call this the peacock-tail. The caudals tend to have a rather dumb-bell shape and are held more or less parallel to each other rather than at the more customary angle of about 60°. Between the lines of insertion of the caudal fins there is an area of the caudal peduncle which carries several scales. In some individuals at times the bulk of the caudal fin may be strongly flaring and held at right angles to the body. Viewed from behind, the lobes of the caudals take up an X-like (or St Andrew's cross) configuration which provides the basis for one of the common Japanese names for the variety, Kujyakuwo or peacock-tail.

Although it is a very striking variety indeed, the Jikin has never become abundant or commonplace. The heritability of the distinctive caudal peduncle and fins is not very high; Matsui pointed out that at the very best with good brood stock, less than half the progeny would develop the characteristic Jikin tail. This difficulty is compounded by the low frequency of the desired colour and colour pattern in Jikin progeny. It is common knowledge that some of the best Jikins are the result of human artifice. Many cosmetic operations are reputedly carried out involving removal of red scales in the wrong places or the use of chemicals to decolourise them. Current examples of the variety have much paler colour than those which were to be seen in the 1970s.

The Jikin is certainly very much a variety for the dedicated specialist, who would probably seek to improve the heritability of the tail features and colour pattern. They might also seek to improve the dorsal and ventral contours which often are lacking smoothness.

Fig. 4.7 Jikin.

Table 4.5 Description of the Jikin.

Body:	Elongated, laterally compressed, depth 3/8–1/2 body length.
Fins:	Dorsal fin single, anal, pectoral and pelvics paired, caudal fins double, distinctive peacock-tail conformation, caudals dumb-bell shaped, strongly flared from caudal peduncle. Spread of caudals up to 1.5 times body depth. Dorsal height 3/8–1/2 body depth.
Eyes:	Normal.
Contours:	Smooth with approximately equal dorsal and ventral curvature and smooth lateral contours, there is a tendency towards angular ventral contours.
Coloration:	Very distinctive body, silver (or white) with red lower abdomen, finnage deep red. The status of the calico Edo-jikin is uncertain.

The Fantail

The Fantail (Fig. 4.8; Table 4.6) can be regarded as a development of the Wakin which epitomises the kind of change which has become characteristic of the more highly evolved fancy goldfish. The body is shorter with a tendency to become somewhat egg-shaped and the fins are appreciably longer. The dorsal and ventral contours are smooth and almost symmetrical. The Fantail is a deservedly popular fish and it is colourful and hardy, suitable for both aquarium and pond. It occurs in metallic and calico forms, the metallics are probably somewhat hardier. There is no reason why any of the standard metallic colour variants should not be produced but the commonest are the red and bicoloured red and white metallics, followed by the calicos which are popular enough but not produced in anything like the same quantity.

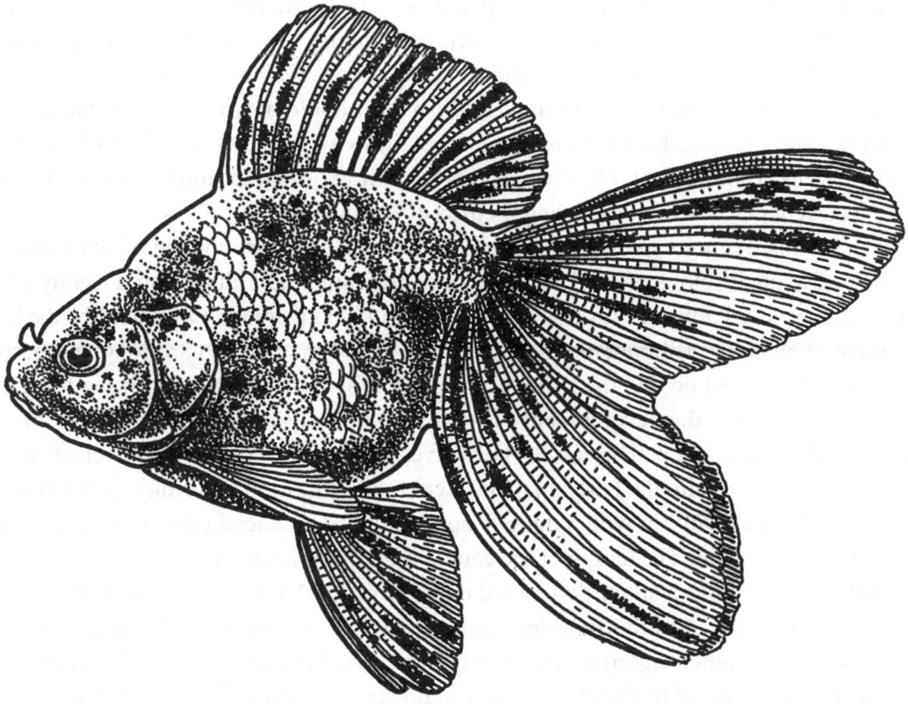

Fig. 4.8 Calico Fantail.

Table 4.6 Description of the Fantail.

Body:	Short, globular, depth greater than 3/5 body length.
Fins:	Dorsal fin single, anals, pectorals and pelvics paired, caudal double, bifurcated held without drooping below horizontal (as in Wakin), spread up to 1.5 times body depth, length of caudal half body length with rounded lobes, dorsal 1/3–1/2 body length.
Eyes:	Normal.
Contours:	Smooth dorsal, ventral and lateral contours.
Coloration:	Both metallic and calico forms are produced.

The Fantail can be considered as something of a halfway house between the Wakin and the Veiltail. Something very much like the modern Fantail must have been the precursor of the more highly developed Ryukins and Veiltails. The Japanese Watonai, produced from a Ryukin × Wakin cross is in effect a reconstruction of the Fantail. Since the end of World War II there have been changes brought about in the British Fantail. Relatively shorter finnage has been retained but substantially greater body depth is commonly seen. The plane of the caudal peduncle should be horizontal and the caudals be held in the same line. It is quite important that the distinctive characteristics of the Fantail be maintained and that it does not become a ragbag of sub-standard material culled from other varieties.

The Ryukin

The Ryukin or Japanese Fantail (Fig. 4.9; Table 4.7) is an exceedingly heterogeneous variety. This probably arises from its status in Japan where it is certainly the most popular fancy goldfish kept in that country. In the early twentieth century it was known as the Fringetail, not the most appropriate name for a goldfish variety but one which could usefully be used to denote this rather diverse group of fancy goldfish. We should bear in mind the fact that the Veiltail was originally called the Japanese Fringetail, so we must not lose sight of the affinities of the Ryukin and Veiltail. These two forms are now very distinct and few would disagree with the current practice of considering them as different varieties.

Matsui in his writing expressed disapproval of the rather short finnage which was characteristic of many Ryukins offered for sale; these offending individuals could rightly be regarded as Japanese Fantails. Presumably what Matsui would approve of would be the lines which could be called Ribbontails. In these the lobes of the caudals are long and narrow and tend to stream out behind the fish when it moves forward. The most striking feature of the modern Ryukin is the development of what is commonly called 'the hump'. In mature Ryukins with well-developed humps the head gives the impression of being attached, almost as an afterthought to a circular body. In marked contrast to the modern Ryukin is the example depicted by Smith (1909) in which there is no strongly pronounced concavity between the snout and the dorsal fin. The length of the caudal is approximately equal to that of the body, the caudal peduncle has a slight downward curve while in the modern Ryukin this may be more or less horizontal. Smith's type does provide a link of sorts between the modern Ryukin and the Veiltail. Another important contrast between Ryukin and Veiltail is the form of the caudals. The broadtail of the Veiltail came into favour in its early development and the so-called 'Swallow-tailed Veiltail' mentioned in some pre-war literature is probably what now

Fig. 4.9 Ryukin.

might be called the 'Fringetail'. It is perhaps fair to conclude that the major feature of the Ryukin of today is the great depth of body, which may be over 80% of the body length. The old Fringetail line had clearly great potential for differentiation into several distinctly different

Table 4.7 Description of the Ryukin.

Body:	Short and globular, depth 3/4 to more than body length.
Fins:	Dorsal fin single, anals, pectorals and pelvics paired, caudal fin double, strongly bifurcated with downturned carriage length 3/4 to 1.5 times body length with rounded lobes, spread of lobes may exceed body length, dorsal height 1/3–1/2 body depth.
Eyes:	Normal.
Contours:	Dorsal contour shows characteristic hump, ventral and lateral contours smooth.
Coloration:	Metallic, red, orange, red and white and white individuals are common, as are calico variants.

types which might vary with regard to finnage on the one hand and body form on the other. The proverbially rugged constitution of the Ryukin would be a great help in realising these objectives.

A wide range of colour variation is possible in the Ryukin. Pure red and red and white metallic are most commonly seen, calico Ryukins are deservedly quite popular but supply probably is not adequate to meet potential demand. Calico Ryukins have been developed with attractive coloration; however, development of body depth is less extreme. (Other metallic colours could be produced without undue difficulty should the demand develop for them.)

The Tosakin

The Tosakin or Goldfish of Tosa (Fig. 4.10; Table 4.8; Plate 5) arose in the vicinity of Kochi City, as Tosa is now named. There are two possible origins of this: the earlier is that it arose from a cross between a Ryukin and an Osaka Ranchu in the late Edo period prior to 1868, the later is by direct mutation from the Ryukin as reported by Matsui. The distinctive feature of the Tosakin is its tail which is essentially a web-tail with a very characteristic shape. The free (i.e. the lower) lobes of the caudal are extended laterally and curl forwards. In an outstandingly good specimen the tail configuration is very elegant indeed. Unfortunately the Tosakin is not an easy fish to breed, rear or keep. Usually it is kept in shallow water and the tail conformation is thought to produce so much drag in swimming that natural spawning can be difficult and it may be necessary to hand spawn. It is difficult to understand quite why the Tosakin should be such a difficult subject. The body conformation is Ryukin-like but less deep and it seems therefore most probable that the mutation producing the caudal conformation has deleterious pleiotropic effects. The fish created an enormous stir in the 1970s when introduced to the West but nowadays it is rarely seen even in Japan.

Table 4.8 Description of the Tosakin.

Body:	Short and globular, depth 2/3 or more of body length.
Fins:	Dorsal fin single, anals, pectorals and pelvics paired, caudal fin web-tail, with characteristic flowing conformation, length approximately 2/3 body length, dorsal height 1/3–1/2 body depth.
Eyes:	Normal.
Contours:	Dorsal contour may show slight hump, ventral and lateral contours smooth.
Coloration:	Metallic, red, orange and red and white are commonest.

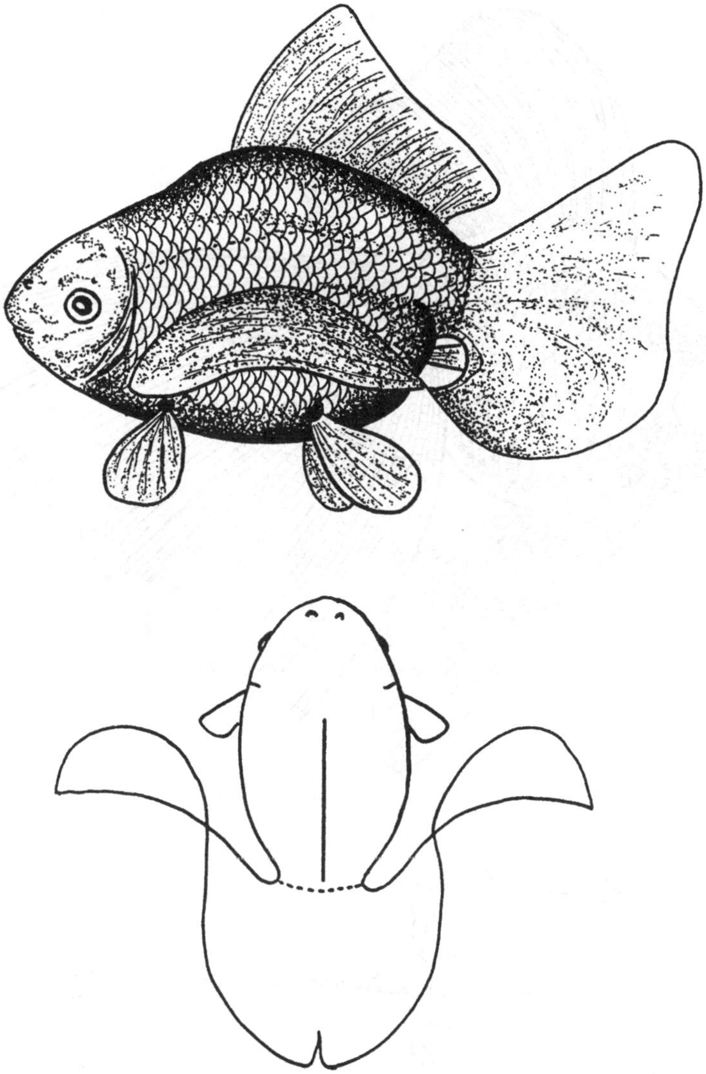

Fig. 4.10 Tosakin.

The Veiltail

The Veiltail (Fig. 4.11; Table 4.9) is a truly remarkable fish and in the view of many Western goldfish fanciers is the ultimate goldfish. This esteem is not shared in the Far East, in much the same way that Western Shubunkins have not made any impression there either. Perhaps part of the reason for this is that neither the Veiltail nor the Bristol Shubunkin has been exploited commercially. If commercial production has been attempted then to date it has not been successful; both the Veiltail and Bristol Shubunkin are predominantly maintained by the efforts of amateur breeders. While they may come on the market, it is only on a small scale and from a very limited number of producers.

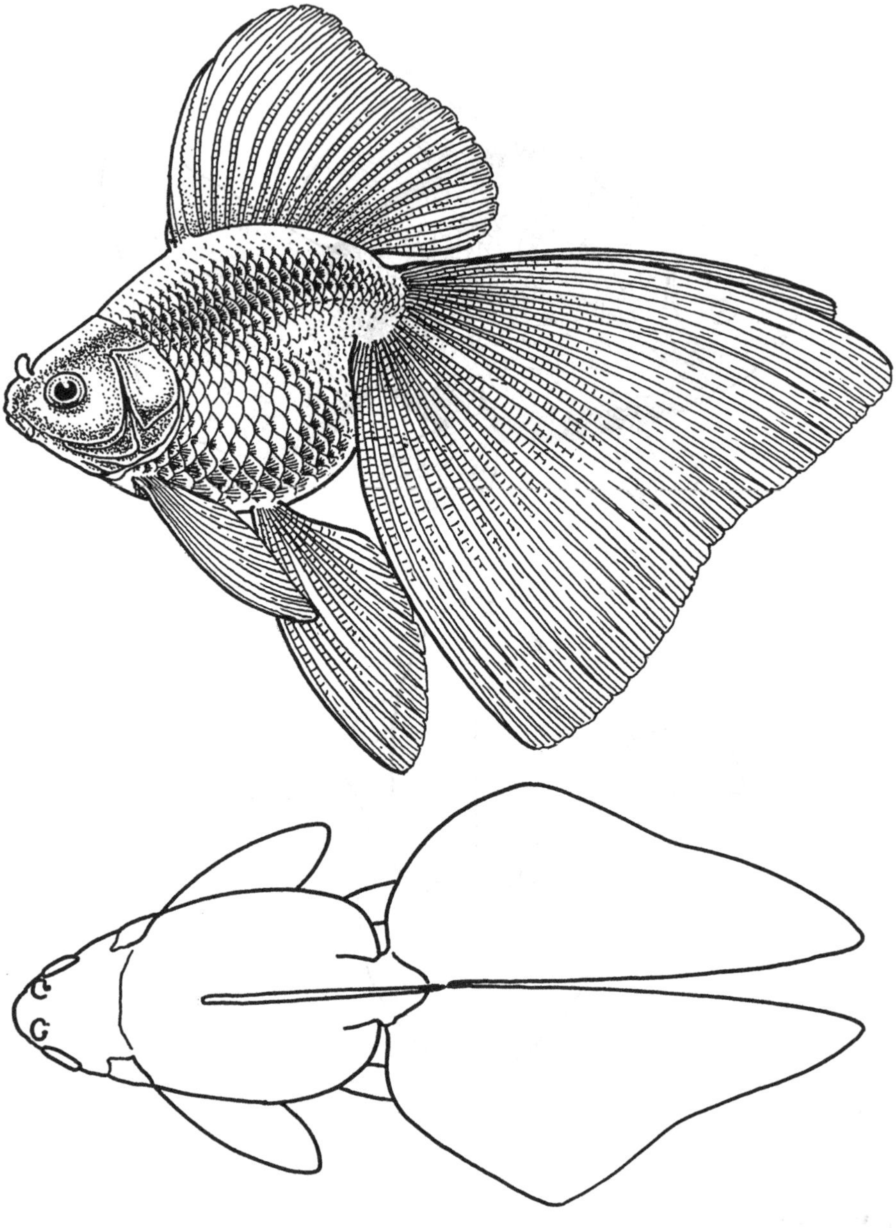

Fig. 4.11 Veiltail.

Perhaps the most curious feature about the Veiltail is that it came upon the scene in almost its definitive form in the early years of the last century. The Veiltail illustration in Wolf's (1908) work still represents very largely the ideal; the only change of substance is that the tail depicted is slightly bifurcated while the current standard is for the broadtail with little or no indentation of the trailing edges of the caudals. Whereas in most other varieties of goldfish the passage of the twentieth century has seen many very substantial changes and developments, the changes in the Veiltail on the whole have been relatively minor. The most significant has

Table 4.9 Description of the Veiltail.

Body:	Short and globular, depth 3/4 or more of body length.
Fins:	Dorsal fin single, anals, pectorals and pelvics paired, caudal fin double with straight-cut trailing edge, very broad with a downward carriage, length 3/4–1.5 times body length. The dorsal fin should equal or exceed the body depth.
Eyes:	Normal.
Contours:	Dorsal contour smooth without a hump, ventral and lateral contours smooth.
Coloration:	Metallic and calico variants have been produced.

been the production of the calico form of Veiltail. Although the standard was set early in its history, it has not been an easy one to maintain. There are some truly outstanding red metallic Veiltails to be seen at the present time but the same cannot be said for the calicos. It is a very great challenge to produce quality Calico Veiltails and since the end of World War II many members of the GSGB have devoted considerable efforts to the production and maintenance of high quality Calico Veiltails; it has been an uphill struggle and with the advancing age and declining energies of the 'old guard', the future is uncertain.

It is worth examining the essential characteristics of the Veiltail as exemplified by the best current examples of red metallic Veiltails. In a mature fish the depth of body will be 75–80% of the length, the height of the dorsal will be equal to the depth of the body and it will be full and sail-like, the trailing edge approximately horizontal. The caudals should be one and a half times the length of the body and drape gracefully downwards in a gentle curve. The line of this curvature should be with a gentle transition from that of the dorsal contour. These specifications are met without undue difficulty, as already indicated, in metallics but there is a problem in producing calico individuals to the same standard. This arises from a pleiotropic effect of the mutant gene producing transparency of the scales in producing softer finnage and less robust and strong fin rays. Selection for fin length may be successful but such fish at maturity have more finnage than they can cope with. The leading dorsal fin-rays may bend backwards or sideways and the appropriate rays of the caudal fins may lack the strength to produce the required graceful carriage. This could result in a disastrous collapse of the major fins and total loss of the characteristic elegant deportment of the good Veiltail. The breeder of Calico Veiltails has therefore had to select less strongly for length of finnage and exercise a fine judgement in doing so to avoid the over-finnage problem. Another problem is the loss of strength of colour in calicos. In raising some metallic scaled progeny from Calico Veiltail spawnings, occasionally some individuals undergo demelanisation. If this were to occur in a Calico Veiltail it would result in total loss of the desirable blue coloration. A further problem that has attracted the attention of some American Veiltail breeders is the difficulty that particularly deep-bodied Veiltails experience in controlling their buoyancy. This can be due to distortion of the swim bladder arising from the shortening of the body, or from distension of the alimentary canal (gut) as a result of bacterial fermentation of food and generation of gas. The discharge of the latter from the digestive tract may be slowed down by its distortion in the body cavity arising again from the shortness of the body. As a result some breeders now favour a less deep body conformation.

It is fair to say that the Veiltail and especially the Calico Veiltail has been a constant challenge to Western goldfish breeders for the whole twentieth century. Its origin could be regarded as a sport or off-type produced from Ryukin stock. The Aquarium Society of Philadelphia will always be associated with the Veiltail and its development and establishment as the very pinnacle of perfection in the goldfish world. Recently there has been a spate of anecdotal evidence suggesting that acceptable Veiltails have been selected from among the off-type progeny of Butterfly fish and second-grade imports of Ryukin rejects. This is perplexing but it does suggest that the Veiltails which were kept between the wars in Central and Eastern Europe may have had an independent origin from the Philadelphia stock. Unfortunately these stocks were lost during the Second World War and surviving records concerning them are very scarce.

The Telescope

The name telescope-eye has been used for well over a century to describe the protuberant eyes of some goldfish (Figs 4.12–4.15; Table 4.10). Telescope-eyes have been depicted in Oriental art for over 200 years. The name has been hallowed by long usage and, while it can be criticised as being not entirely appropriate, the attempt to substitute the more reasonable 'globe-eye' has not been successful. *Vox populi* favours the name telescope and this verdict perforce should be accepted.

The condition of exophthalmia which the telescope eye shows can occur in nature as a pathological condition but in the goldfish the cause is genetic and is produced by a simple gene mutation. As shown on the Billardon de Sauvigny scroll, telescope eyes were to be found on both long and short-bodied fish. At the present time the feature is virtually confined to short-bodied fish but fish with telescope eyes collectively show an enormous range of varia-

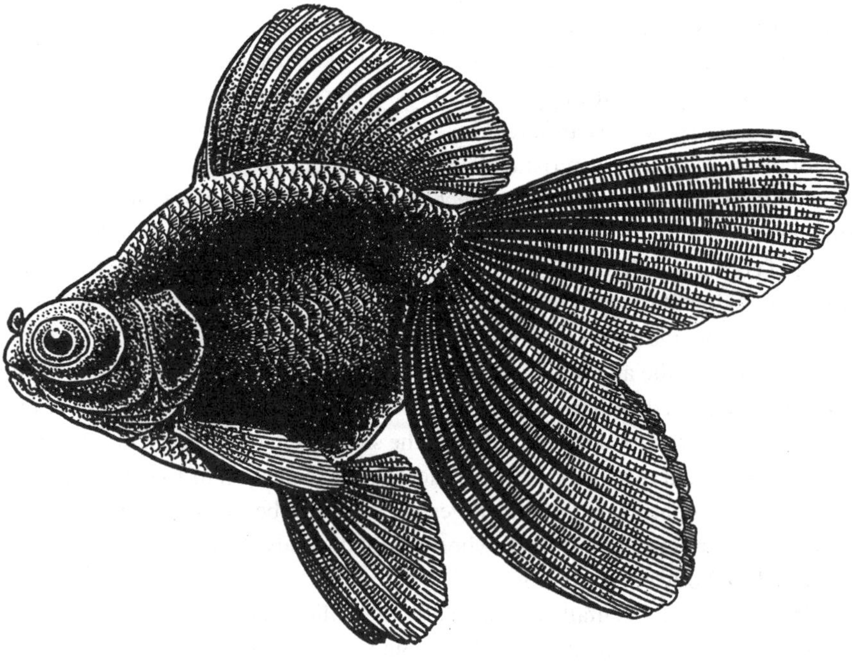

Fig. 4.12 Moor.

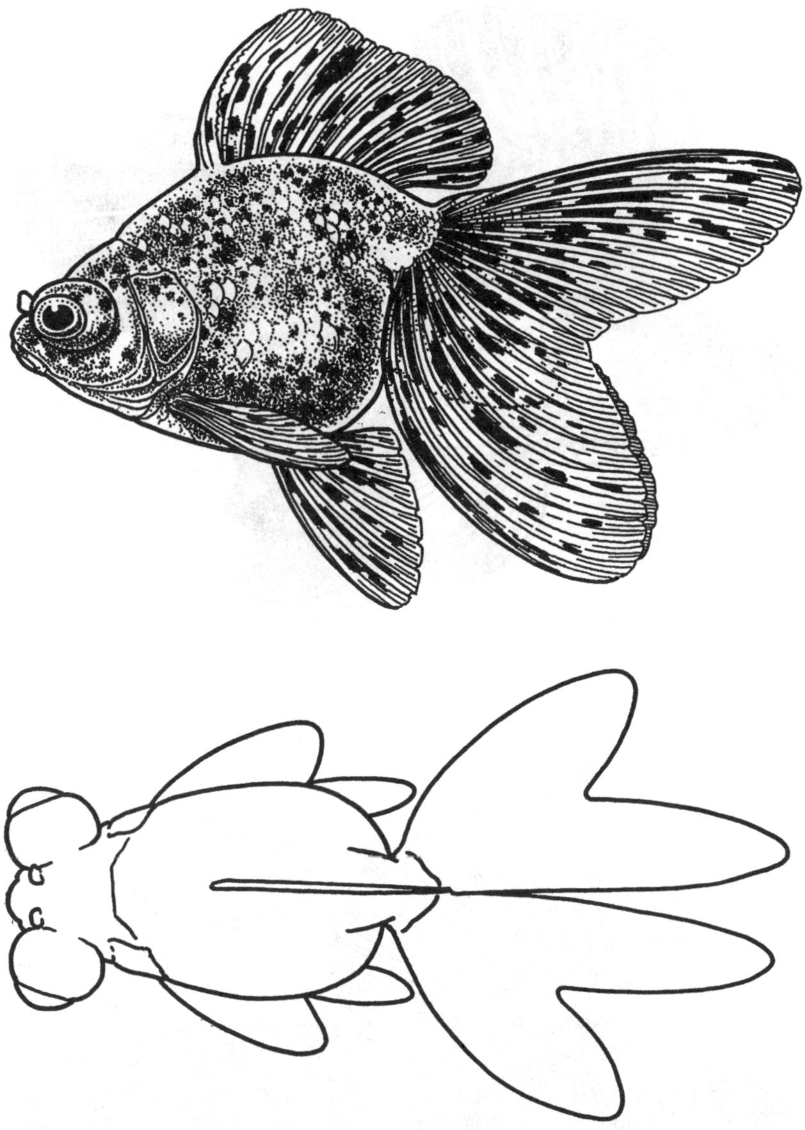

Fig. 4.13 Calico Telescope.

tion. The shape of the eye varies, it may be spherical, ovoid, a truncated cone or a segmented sphere (Figs 2.1, 3.5). In the GSGB standards for Globe-eye and Moor, different eye forms are indicated in the drawings, conical and spherical respectively. The advisability of making such fine distinctions is perhaps a moot point.

The most remarkable feature of this group is the pre-eminence achieved by one type within it. Until the recent arrival on the market of the Butterfly-tails, the vast majority of telescopes seen both at shows and in the shops were Moors (Fig. 4.12), that is, melanics. The very wide range of colour forms, various metallics and even the remarkable and significant Calico Telescope were scarcely to be seen. In the GSGB standards for metallic fish, black is the only

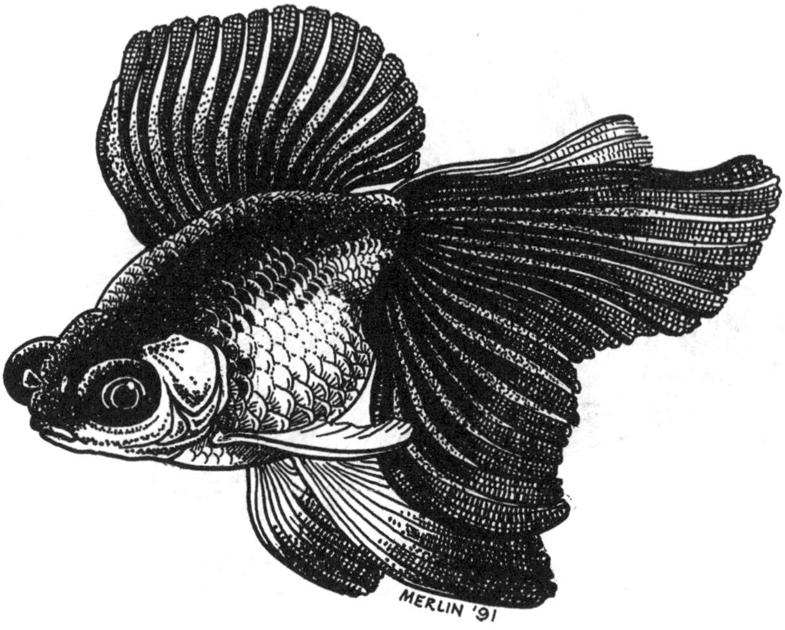

Fig. 4.14 Magpie.

Fig. 4.15 Butterfly tail.

Table 4.10 Description of the Telescope-eye.

Body:	Short and globular, depth 3/4 or more of body length.
Fins:	Dorsal fin single, anals, pectorals and pelvics paired, caudal fin double with variable length, conformation and carriage, this may be similar to the Fantail, Ryukin or Veiltail, in butterfly-tails caudals are held with a near horizontal spread.
Eyes:	Protuberant, may be spherical, segmented spherical, ovoid or in the form of a truncated cone.
Contours:	Smooth dorsal, ventral and lateral contours, a Ryukin-like hump may develop in some individuals.
Coloration:	A remarkable variety of colours are developed, mostly metallics although calicos are known, black is most popular, variegated black and white or red forms have also been produced.

acceptable colour for both Moor and Globe-eye, and calicos are only acceptable for entry in the Globe-eye class. The main difference between these two forms is in the form of the tail, bifurcated caudals in the Globe-eye and a straight trailing edge in the Broadtail. In both, however, the tail fin must be long, preferably as long as the body or longer. In essence the Broadtail Moor is a black metallic Veiltail with telescope eyes while the Globe-eye is what might in the old days have been called a black metallic Fringetail. The enthusiast scene is vastly different from the commercial where virtually the entire offering for sale is of what can best be described as Fantail Moors.

The coloration of the Moor is an interesting and complex subject. The ideal coloration and quality is of a deep velvety black. Although the fish has metallic scales these are so overlain with melanophores and melanin that no metallic sheen is apparent. When such a sheen can be seen, the quality of the colour suffers and it looks dark brown rather than black. This may be the prelude to a colour change and loss of melanin: the conversion of a Kuro-demekin to an Aka-demekin, as the Japanese would say. The attainment of the prized colour of the Moor has been studied by biochemists, who have found higher concentrations of the enzyme tyrosinase in the Moor as compared with other goldfish varieties. Tyrosinase acts upon the amino-acid tyrosine to produce melanin. One presumes that provided the fish receives an adequate content of tyrosine in the diet and the high level of tyrosinase is maintained, then melanin should be produced in abundance and the desirable deep black colour produced. There is, however, an environmental factor to be considered in the equation and that is the quality of light received. It is a matter of common observation that fish receiving the optimum level of natural light develop a wonderful intensity of colour, be they Common Goldfish, Shubunkins or Moors. When the latter are kept indoors continuously then serious consideration should be given to providing supplementary lighting to reproduce as far as possible the natural light spectrum.

It is regrettable that the Calico Telescope (Sanshoku demekin; Fig. 4.13) has apparently fallen out of favour. This variety was a parent of the Shubunkin and the non-recurrent parent of all calico goldfish varieties. This loss of popularity is difficult to understand as it can be a very attractive fish; at the present time it is difficult to obtain examples of the variety. This situation is in strong contrast to the recent upsurge in interest and production of the Pandas, Magpies and Butterflies. There has been something of a problem here in their naming. I have heard the suggestion that the Chinese who have developed all these types call the red and black variegated fish Pandas, and the black and white ones Magpies (Fig. 4.14). The Chinese are familiar with two pandas, the red panda (after which the goldfish are named) and the giant panda; we in the West know the giant panda but very few are aware of the existence of the other. It may seem pedantic to point this out but in the end *vox populi* will have its way and the names 'Panda' and 'Magpie' will probably be synonymous. These two variegated forms are of particular interest; the black and white fish created a sensation when they first appeared and were eagerly snapped up. Unfortunately it seems to be a difficult variety to keep, breed and reproduce the quality which was so very impressive in the original introductions. The red and black form is also a striking fish but is also of particular interest in that what has been regarded as a transitional (but beautiful!) colour phase has been stabilised, to a degree at least. No doubt further selection will enhance stability and colour quality. One of the great qualities of Chinese goldfish breeders is that they are very open-minded and recognise opportunities to develop new variants, compared with Western breeders who tend to be blinkered by standards. I myself have observed Moors that in maturity underwent a colour change, where some individuals retained the intermediate red and black mixture of colours for a number of years; the colour change did not go to completion. In another instance a spawning of commercial Fantail Moors produced an appreciable number of fish which held their caudals in a more or less horizontal plane; they were in fact butterfly-tails!

One of the disappointing features of many introductions of new goldfish varieties to this country is that they have for a variety of reasons failed to establish an enthusiastic band of devotees. This has often been because they have not survived where they have been introduced. Many Pandas which came into the possession of skilled and experienced breeders failed to survive for longer than six months and when any survivors were bred the results were disappointing. If it is any comfort to Western breeders they are not alone in this experience; it is recognised by Chinese breeders that successful breeding of some varieties can only be achieved in specific locations. The reasons for this are anything but clear except that there must be environmental effects to do with local climate and water quality.

The distinctive feature of the Butterfly (Fig. 4.15) is the tail which is held with the twin caudals spreading horizontally. In essence it is a relatively minor variation on the Telescope-eye theme; the refreshing feature is that many of the characters of the old Telescopes such as red metallics and calicos could make something of a comeback in Butterfly-tail guise. Another character which is apparent in modern Chinese Telescopes is in the greater size of the eye in some types. A concentric ring of a contrasting colour may also be seen circling the iris on the telescope eye. Some successful attempts have been made to produce interesting-looking silver and matt fish ('pinkies') with telescope eyes and red heads.

What has happened in the last few decades in China is an object lesson in what can be accomplished when creative perception and imagination is given its head. The range of new variants produced in China recently is absolutely breathtaking. The Chinese have certainly re-established their pre-eminence in the goldfish world with a vengeance.

The development of the globe-eye feature is interesting. Initially the eyes are indistinguishable from the normal or wild-type. Progressively they become protuberant during the course of the first year. Development of the two eyes is not necessarily equal or symmetrical. One eye may be protuberant, the other not, one may be larger than the other even to the extent that one of the eyes may even be smaller than normal while the other shows enlargement and protrusion in the typical fashion. Naturally the breeder will cull such off-types, which will thus tend to become progressively less frequent in subsequent generations.

The Celestial

The Celestial (Fig. 4.16; Table 4.11) is in essence a Telescope-eye in which the eye faces skywards rather than laterally. Initially the course of development is as in the Telescope, the eyes enlarge and protrude laterally then the irises of the eyes rotate to face forward and this is followed by further rotation in which the iris comes to face upwards. This development is subject to the same vagaries as that of the Telescope-eye but in addition the rotation processes may not be followed equally, or they may not occur at all, which results in a Telescope-eye the size of which may be appreciably larger than is customary. The eyes may face forward and not turn skyward or there may be a true Celestial eye and a forward-facing eye on the same individual. It is possible that this development may be temperature sensitive; progeny from

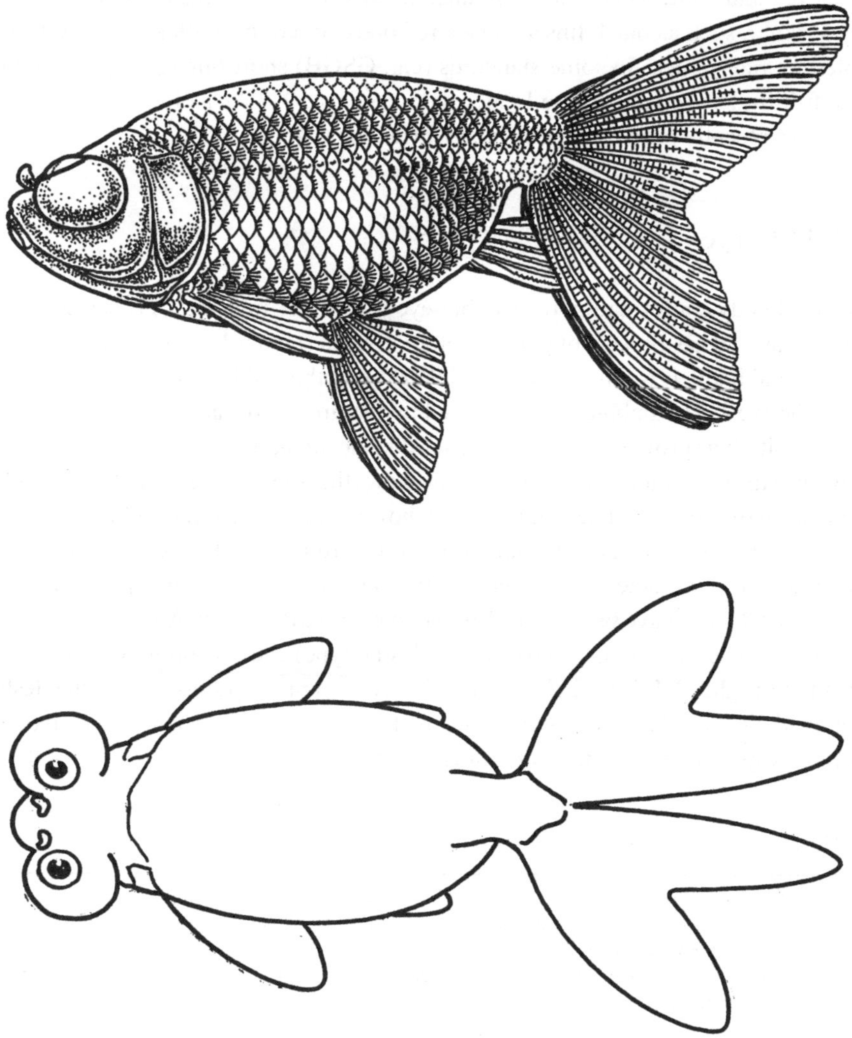

Fig. 4.16 Celestial.

Table 4.11 Description of the Celestial.

Body:	Short and globular or medium length and less globular, depth 1/2–3/4 body length.
Fins:	Dorsal absent, anals, pectorals and pelvics paired, medium-short in length, caudal fin double 3/8–1/2 body length, bifurcated.
Eyes:	Large, protuberant, upwardly directed.
Contours:	Dorsal, ventral and lateral contours smooth.
Coloration:	Common metallic colours and calicoes are found.

late spawnings raised at ambient temperatures may be more likely to experience incomplete development of this special characteristic.

The absence of the dorsal fin is a characteristic feature of the Celestial and, although the standards may specify a short body, the majority of Celestials tend to have a body length intermediate between a Wakin and a Ryukin. The Celestial is a metallic fish commonly red-orange or red and white in colour. Individuals with strongly contrasting coloured fins can be very striking, such as black fins against a red body or red fins with a silver body. Calico examples are rarely seen. In some standards (e.g. GSGB) short finnage is preferred but in Chinese fish fin length tends to be longer with caudals 50–75% as long as the body. The Chinese have recently produced a Celestial with a dorsal fin.

The Bubble-eye

In its general body conformation the Bubble-eye (Fig. 4.17; Table 4.12) is basically similar to the Celestial and has finnage of similar length. British standards as in the Celestial favour shorter fins and body. It is only since World War II that the Bubble-eye has been seen commonly in the West. The bubbles are produced by the growth of large vesicles arising from the lower orbit; their growth frequently causes displacement of the eye so that the iris faces upwards as in the Celestial. Close examination shows that the eye itself is not in fact enlarged in the majority of such fish. The Chinese have, however, combined the Bubble-eye with the Celestial eye and in such Celestial Bubble-eyes the protuberant Celestial eyes are present combined with the characteristic vesicles of the Bubble-eye. Although typically lacking a dorsal fin, the Chinese have produced a Bubble-eye with a dorsal fin. A form of Bubble with rather small vesicles and upturned (but not Celestial type) eyes is commonly known as the Toadhead or Froghead. It is probably more often seen in photographs than in the flesh. The bubbles are delicate and easily ruptured and great care is required in handling to avoid damage; they may grow to a centimetre or more in diameter.

Table 4.12 Description of the Bubble-eye.

Body:	Short and globular or medium and less globular, depth 1/2–3/4 body length.
Fins:	Dorsal absent, anals, pectorals and pelvics paired, medium-short in length, caudal fin double 3/8–1/2 body length, bifurcated.
Eyes:	Characteristic 'water bubble-eye' developed, an infraorbital vesicle, eye may be displaced with iris directed upwards.
Contours:	Dorsal, ventral and lateral contours smooth.
Coloration:	Common metallic colours and calicoes are found.

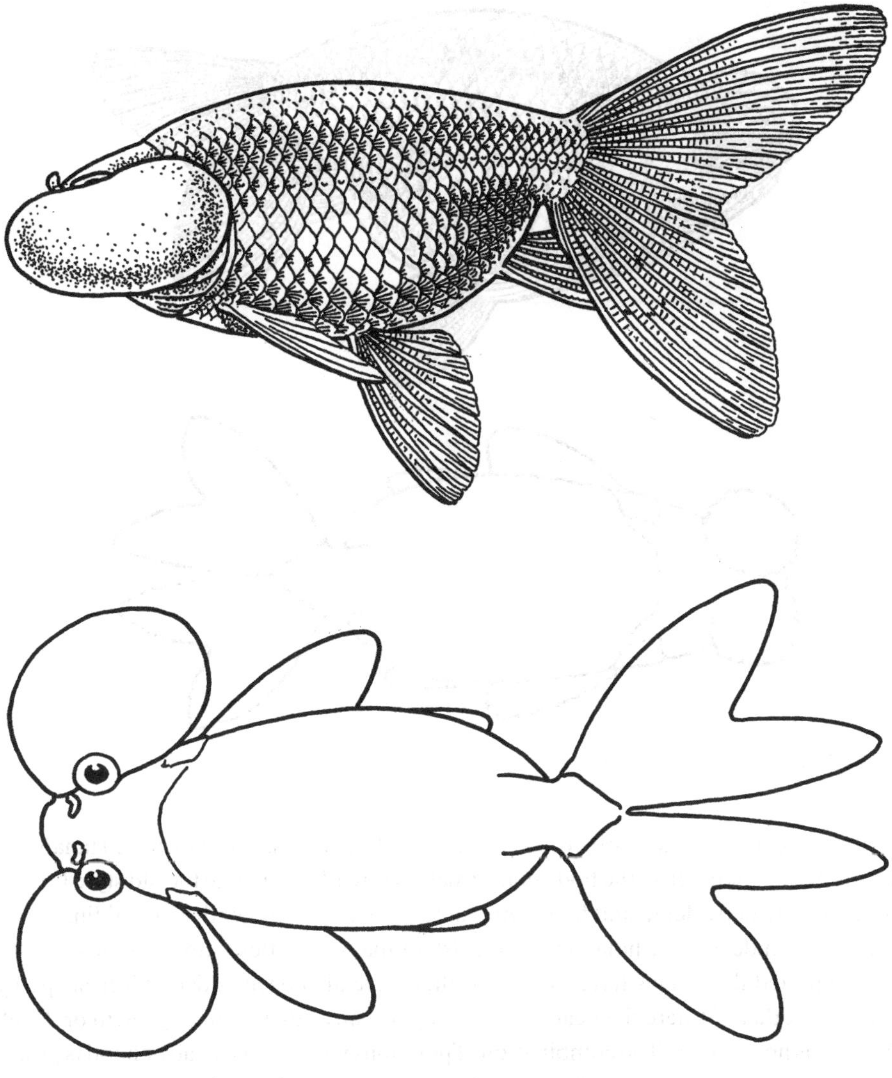

Fig. 4.17 Bubble-eye.

The usual metallic colours are to be seen in Bubble-eyes – reds, red and white, orange and black – and in recent years some particularly striking calico examples have been seen.

The Pompon

There are two forms of the Pompon (Fig. 4.18; Table 4.13), the form commonest in the West is dorsal-less but that favoured in Japan has one and is known there as the Hanafusa. According to British standards, body form and finnage are as in the Celestial and Bubble-eye. The distinctive feature is the pompon produced by the hypertrophy of the nasal septa which produce ball-like appendages on the top of the head. Ideally these should be spherical in shape

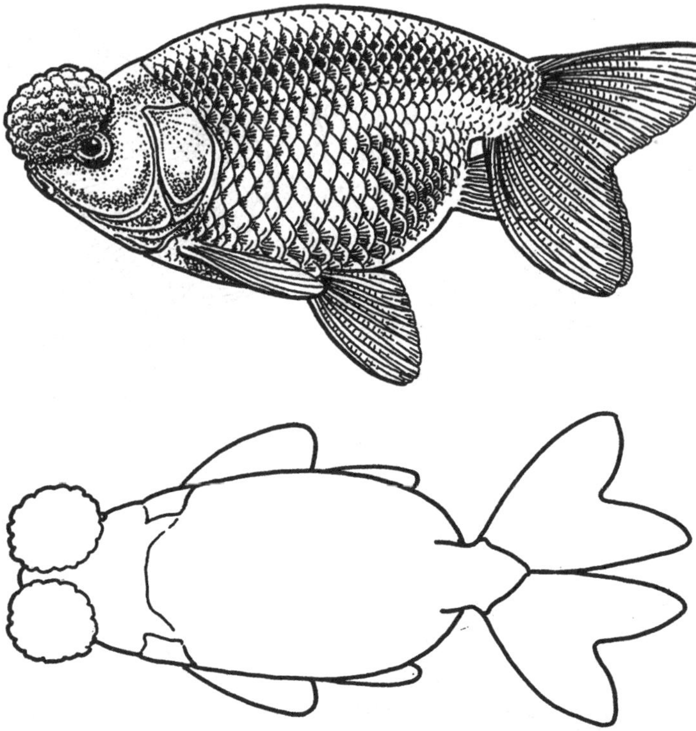

Fig. 4.18 Pompon.

and firm in texture; they are sometimes called narial bouquets. The Japanese Hanafusa is a rather different looking fish: the body conformation is really more like the old style Fringetail with a moderate body depth and fin length, rather than a Pompon with a dorsal fin.

There is a tendency in China to produce fish of many varieties with pompons. Most Celestials imported these days have pompons; they have also been added to Telescope-eyes, Orandas and others. Where the head does not support any other form of growth or development, there is no problem. The combination of pompons with a goosehead or hood is probably counter-productive as neither feature may then be seen to the best advantage.

Table 4.13 Description of the Pompon.

Body:	Short and globular or medium and less globular, depth 1/2–3/4 body length.
Fins:	Dorsal absent, anals, pectorals and pelvics paired, medium-short in length, caudal fin double 3/8–1/2 body length, bifurcated.
Eyes:	Normal.
Contours:	Pompons (narial bouquets) developed on dorsal contour, dorsal, ventral and lateral contours smooth.
Coloration:	Common metallic colours and calicoes are found.

The Pearlscale

The Pearlscale (Fig. 4.19; Table 4.14) is a very distinctive variety characterised by the eponymous scale which is convex rather than flat with a conspicuous centrally placed boss – the pearl. The body proportions in relation to the finnage are basically those of the Fantail; however, the body characteristically is very globose, almost as though the fish was dropsical. The caudal fins are usually indented but variants occur where this is straight-cut and the fins generally are longer. The Pearlscale is quite popular and commonly seen at shows and in the shops. In the Far East and especially China the pearlscale characteristic has been introduced into many other varieties. It is probably only a matter of time before pearlscale variants of every major variety of goldfish are produced. How many of these will persist is another matter. The Hamanishiki is basically a Pearlscale Oranda in which the hood has produced a paired vesicle on the crown of the head which is a very distinctive feature. It has become so popular in China that the original Pearlscale has all but disappeared.

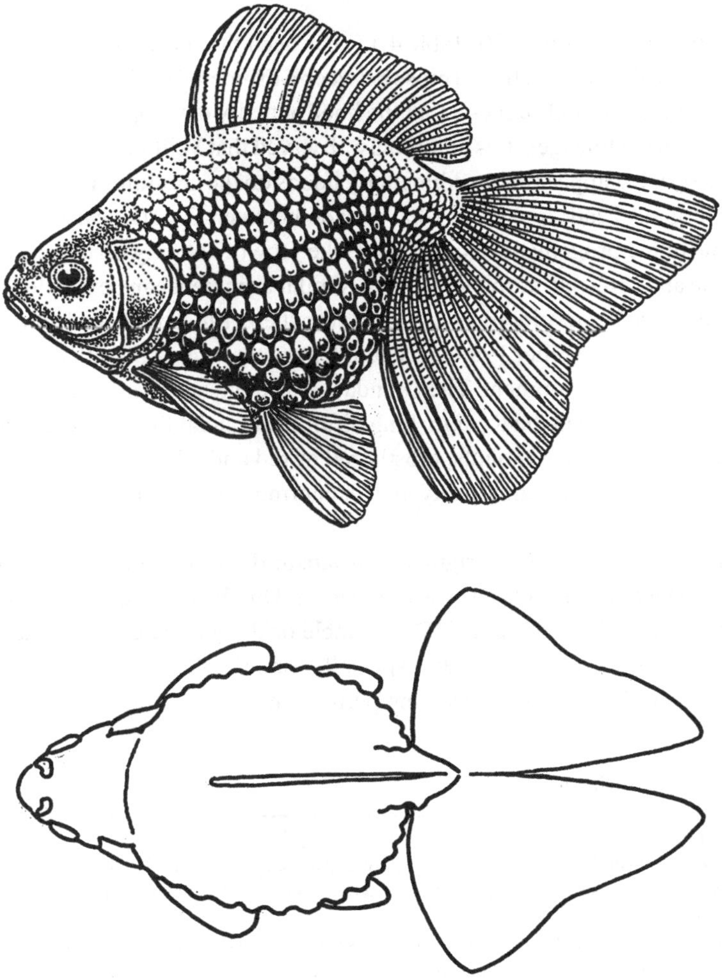

Fig. 4.19 Pearlscale.

Table 4.14 Description of the Pearlscale.

Body:	Short and extremely globular, body depth greater than 2/3 length, carries characteristic domed scales.
Fins:	Dorsal single, anals, pectorals and pelvics double, caudal fin double, variable in shape, forked or square-cut, and in length 1/2 that of body or longer.
Eyes:	Normal.
Contours:	Dorsal, ventral and lateral contours smooth, the trunk is almost circular viewed from above.
Coloration:	A wide range of colours are produced, both metallic and calico.

The Pearlscale is found in a range of metallic forms in addition to the calico. The original form is very appealing and attractive and it would be worthwhile to take appropriate steps to safeguard its future.

The Oranda

The origin of the Oranda (Fig. 4.20; Table 4.15) has been the subject of some controversy. It has been suggested that it arose from a cross between a Fringetail (or Ryukin) and a Lionhead. Matsui has disputed this and suggests that the hood actually developed independently in the Oranda and Lionhead lineages. It is commonly observed that Veiltails and other non-hooded varieties develop what appear to be rudimentary hoods on the cranium. This could be viewed as an incipient goosehead. In a cross I have made between Crucian carp and red metallic Veiltails, some of the progeny developed hood-like growth not only on the cranium but also on the opercula and in the infraorbital region. A similar argument can be developed regarding suppression of dorsal fin development, as noted elsewhere. Most published goldfish genealogies (Matsui 1972, Teichfischer 1994) accept independent origins of both the hood and suppression of dorsal fin development. The Oranda was established over a century ago as a fish with less depth of body than the Ryukin, long bifurcated caudal fins, a goosehead and a limited range of metallic colour variants, mostly red or red and white. In the past 50 years the situation has changed most dramatically and an enormous range of novel variants has been produced.

It is probable that the Oranda originated in Japan; the full name Oranda Shishigashira means literally Dutch Lionhead. The use of the term 'Dutch' in this context is rather similar to English usage as in Dutch courage, Dutch uncle or double Dutch and certainly implied no actual connection with Holland. Two especially remarkable developments can be noted, namely the Redcap (Fig. 4.21) representing perhaps the ultimate in the goosehead lineage,

Table 4.15 Description of the Oranda.

Body:	Short and globular, depth 2/3 or more of body length.
Fins:	Dorsal fin single, anals, pectorals and pelvics paired, caudal fin double, forked or broadtail with a downward sweep, 3/4–1.5 times body length, dorsal fin 1/2–7/8 body depth.
Eyes:	Normal.
Contours:	Characteristic hood in head region, otherwise smooth dorsal, ventral and lateral contours.
Coloration:	Wide range of metallic colours and calico, variegated and self-colours, contrasting coloration of hood and body, Redcap, bicoloured and multicoloured hoods can develop.

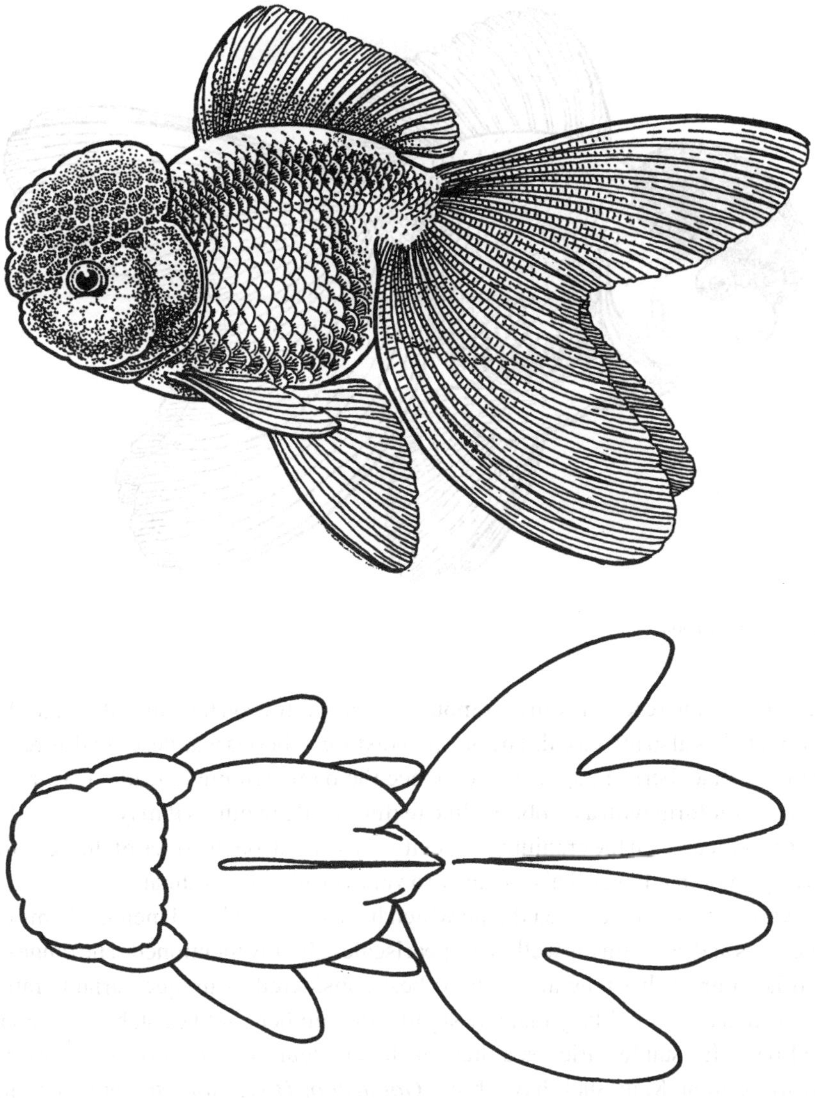

Fig. 4.20 Oranda.

and the Azuma Nishiki or Calico Oranda, whose hood envelopes most of the head. In addition a wide range of colour variants have been produced among the metallics, chocolate blue and black in addition to the customary red, red and white and orange. Some extraordinary variants have been produced in China notably a red-bodied fish with a black hood. Bicoloured hoods, red and white, have also been produced.

There has also been a tendency over the years for the body conformation to become more like that of the Veiltail, deeper body with smooth contours and a square-cut tail or broadtail. There is no doubt that the Oranda can be spectacularly beautiful, it is quite a hardy and undemanding fish and is deservedly popular. Other features which show interesting variation are the form and texture of the hood. The goosehead initially was a relatively compact growth above the level of the eyes; in later developments the volume of the goosehead hood has

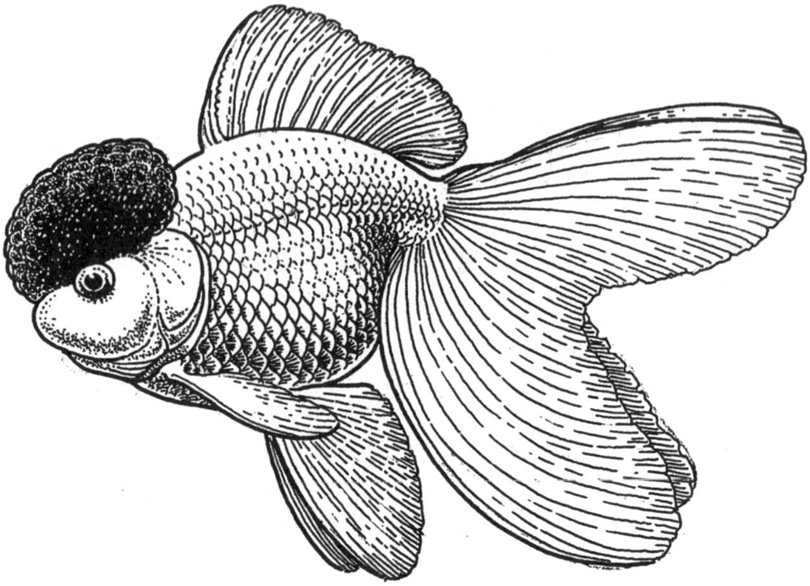

Fig. 4.21 Redcap Oranda.

increased greatly and resembles an enormous red beret sitting on the top of the head. The variation in texture has also increased. The original texture of hoods was described as resembling a raspberry or blackberry in appearance, hence the name 'Bramblehead'. Texture of some hoods may be uniform with a raspberry-like texture while in others it may be quite different. Frequently it is coarse on the cranium and relatively smooth on the sides of the head or it may be relatively smooth all over. This capacity to vary is truly extraordinary.

Other variations on the Oranda theme which have been produced include forms carrying telescope-eyes and pompons as well as the pearlscales already mentioned. The range of forms produced is so great that they are perhaps best considered as unique variants rather than varieties, as it is highly unlikely that any significant number could be stabilised and provide a rational basis for stable varieties either for the amateur enthusiast or in commerce. One has only to examine Man Shek-hay's book *Goldfish of Hong Kong* to appreciate just how enormous is the potential of the goldfish to produce colourful and interesting variants in the hands of gifted breeders. The goldfish fancier may well have to consider taking a leaf out of the Koi breeders' book and both recognise and appreciate fish which far transcend the prescriptive standards we have sought to impose. Standards certainly have and are likely to provide guidelines for the foreseeable future but this may be an appropriate time to think imaginatively on what the future might hold.

The Ranchu-Lionhead Group

The Ranchu-Lionhead group (Figs 4.22, 4.23, 4.24; Table 4.16) is a very complex group of goldfish varieties and even to find an all-inclusive name for the group as a whole is not easy. Matsui's perspective of the group appears to be the most logical. Even the term Ranchu re-

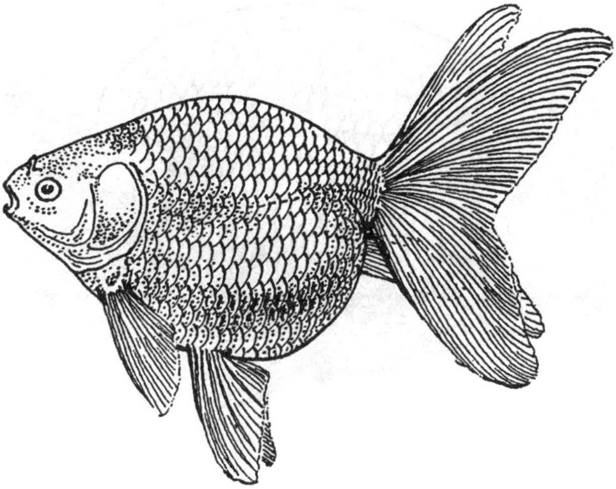

Fig. 4.22 Eggfish.

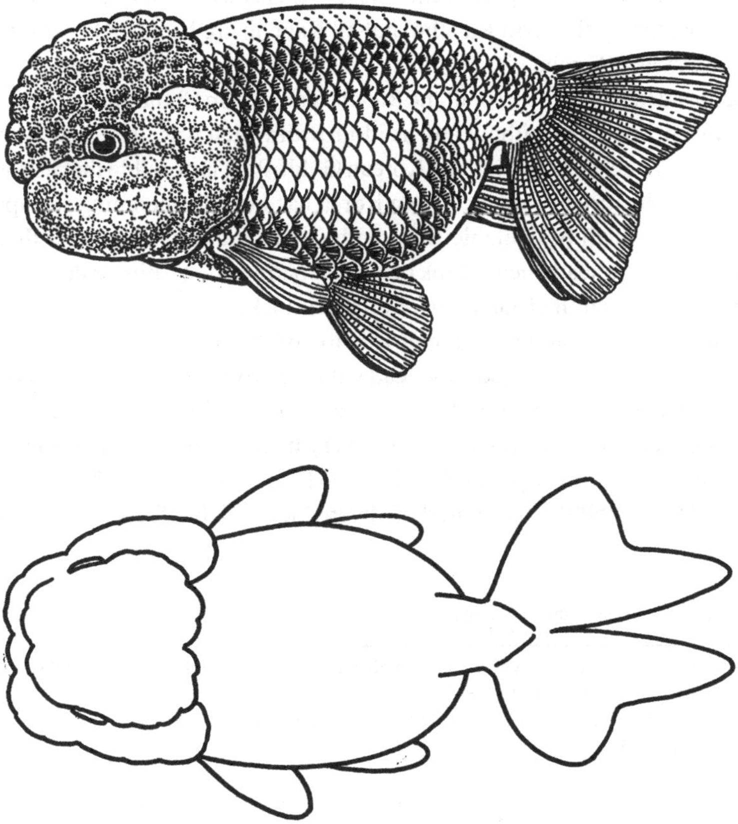

Fig. 4.23 Lionhead.

Fig. 4.24 Ranchu.

quires some qualification as the Celestial has a Japanese name Deme-Ranchu which requires some explanation. Ranchu is a term which in Japanese usage has been applicable to goldfish lacking a dorsal fin; with the great proliferation of variants in the Lionhead lineage the term has been applied specifically to part of the lineage. It is convenient because while all Lionheads can be covered by the term Ranchu, not all Ranchu are Lionheads, strictly speaking, for example the Osaka Ranchu.

Matsui believed that the precursor of the Ranchu came from China and was a short-bodied, twin-tailed, dorsal-less form without any head-growth which was identified as the Maruko. Since the Lionhead has been dubbed 'the Korean Goldfish' it is possible that its route of entry was via Korea. No historical evidence from Korea has been forthcoming in support of this idea. The evidence which is available from both China and Japan suggests that the major development of the Lionhead lineage took place in Japan and that Chinese developments took place following introduction of Japanese developed stock.

Perhaps the best way to attempt a reconstruction of the developmental path of the Lionhead is to compare the fully-fledged Lionhead with the progenitor type, the Maruko or Eggfish (Fig. 4.22). The distinct feature of the Lionhead (Fig. 4.23) is the hood and associated with the development of this characteristic is a very marked broadening of the skull, which in contrast to the rather narrow pointed skull of the Eggfish is very much broader and blunt. In the Lionhead type itself there is a marked divergence which affects the form of the body.

Table 4.16 Description of the Ranchu-Lionhead.

Body:	Short and globular, two types: Ranchu with strongly curved back, Chinese Lionhead with a straighter back, body depth 5/8–3/4 length.
Fins:	Dorsal fin absent, anals, pectorals and pelvics paired, caudal fin double, forked with no downward sweep, 1/4–3/8 of body length.
Eyes:	Normal.
Contours:	Dorsal contour depends on extent of the variable hood development, otherwise dorsal, ventral and lateral contours should be smooth.
Coloration:	A wide range of metallic colours are found and calico, colours variegated or self, body and hood colours similar or contrasting, Redcaps etc., hoods may be bi- or multicoloured.

This has resulted in the establishment of two distinctive forms of body. One of these is not very different from the Eggfish, in which there is a gently curved dorsal contour, with similar caudals; in the other the dorsal contour is more strongly curved and there are also changes in the caudal peduncle and the shape of the caudal fins. The development of this advanced form has been the work of the Japanese and it is to this type (Fig. 4.24) that the term 'Ranchu' is generally applied. The straighter-backed form is often referred to as the Chinese Lionhead, but it is quite apparent from the illustrations in Smith's book on Japanese goldfish that both types were in evidence in Japan in the early twentieth century.

It is perhaps worth noting that many writers on goldfish include the Bubble-eye and Pompon in the Ranchu lineage but not the Celestial. This is quite understandable and emphasises the difficulty in achieving total consistency in the reconstruction of the goldfish family tree. If one accepts that a similar result may arise independently or in different ways, then the situation becomes more amenable to rational analysis.

The development of the hood in the Ranchu lineage seems to have occurred rather differently from the Oranda in that the goosehead hood has been very much less in evidence. A major difference has evolved between what the Chinese call Tigerheads as against Lionheads. In the Tigerhead hood development is uniform with a comparatively smooth surface, while in the Lionhead the three regions, cranial, opercular and infraorbital are recognisable and latterly may show differences in coloration and texture. There are two different types of Redcap Lionheads, one in which has a red goosehead hood, the other in which only the cranial region is red and the others white.

The hoodless Ranchu which are rarely, if ever, seen these days are the Maruko (Eggfish), the Nanking – basically an Eggfish with silver body and red fins – and the Osaka Ranchu which has become virtually extinct, which had no hood but a broad skull and was characterised by a short, globular body, a web-tail and commonly bicoloured. These forms are perhaps of more interest to the goldfish historian than the fancier at the present time.

As in other goldfish varieties the changes have been rung on the Lionhead; examples have been produced with telescope-eyes and pompons. One wonders whether such attempts are really worth the effort. The attempt to produce a Calico Ranchu has not been entirely successful; the colours and patterns have been transferred but hood development as expressed in the Edo-Nishiki has not been impressive. The efforts of American breeders of late have been more successful in producing satisfactory hoods combined with strong coloration. This is yet another example of the goldfish surprising the breeder. One might well ask, as it has been possible to produce Calico Orandas with excellent hoods, why it has been so difficult to do the same with the Lionhead? Mention should also be made of the attempts to produce long-finned Lionheads through crossing with the Oranda while selecting for Lionhead body characters and dorsal-finlessness. A pleasant fish resulted but it made no particular impression and has not found favour. However, this attempt produced the Phoenix, which has aroused some interest. This type exists in a range of variant forms; some have hoods, others are hoodless, they may have metallic or transparent scales, but the common features manifest in the best examples are a rather elongated body of Lionhead conformation but with a very gently curving dorsal contour. The caudals are bifurcated and should be at least as long as the body and preferably longer. When moving slowly by the action of the rather long pectoral fins, the caudals drape gracefully downwards almost vertically; the Japanese name for this can be translated as 'feather dress'. The calico form of the Phoenix was introduced to Britain in the

1970s as the 'Floral Phoenix' and created a strong and favourable impression. Unfortunately further importations were not made and its potential could not be exploited.

Interest in the Ranchu is still strong in Japan and also the United States. Imports of Japanese stock have been made by Chinese breeders who are working their own magic on the variety and the goldfish world will no doubt be surprised and delighted at what emerges in the near future.

Concluding remarks

If one surveys the goldfish scene outside China there are some signs of new and innovative developments in a relatively small number of areas, for example the developments of new variations on the Shubunkin theme in the United States. When one looks at the Chinese scene the range of new variants is overwhelming. Some of this can be regarded in the light of fitting additional bells and whistles to standard models, but beyond the gilding of the lily there is an enormous range of new, beautiful and interesting variation being generated. Reference to three recently published books would serve to emphasise these points. The comparatively recent publication (1993) of Man Shek-hay's *Goldfish in Hong Kong* presents a good sample of the current diversity in Chinese goldfish. This reinforces the message of Li Zhen's earlier (1988) publication *Chinese Goldfish* published in Beijing. The third member of this trio is Bernhard Teichfischer's *Goldfische in Aller Welt* (1994) which provides an interesting European perspective on what is happening in China. There is indeed so much new information to assimilate that it may yet take some time for a considered judgement to emerge on its significance and probable impact.

Unfortunately none of these works is very easy to obtain and it is an added difficulty that Teichfischer's work is not available in English translation. This work includes an informal analysis of the goldfish variation position in China and goes well beyond the mere presentation of information. The reader is strongly urged to secure copies of these works, if only temporarily, to gain familiarity with what is currently emerging from China. Not all of this, as I have indicated, is necessarily to one's liking but most of it will excite wonder and admiration of the sheer genius and creativity of Chinese goldfish breeders.

Chapter 5

The Aesthetic Appreciation of Goldfish

Goldfish appreciation is a complex topic and for convenience it can be considered to embrace not only the qualities which we admire most in our best fish but also those which are embodied in the standards which are set by the Societies for the purpose of judging fish at shows. The breeder naturally interested in producing fish of the highest quality for sale or for the show bench, and in setting selection objectives in breeding programmes, must keep aesthetic considerations well in mind.

Koi fanciers have developed appreciation of their fish to a high level, but it does tend to be a rather esoteric, not to say, arcane business. The lead is very definitely given by the views of enthusiasts and cognoscenti in Japan and the criteria used in the evaluation of Koi range from the eminently sensible to the extremely arbitrary. However, Japanese judges use these criteria as guidelines rather than rigid prescriptions, which leaves an opening for progressive evolution of aesthetic taste, which is as it should be. In the West there is a regrettable tendency to regard show standards as being rigidly prescriptive and leaving the judge only to apply the pre-ordained tariff to the classes before him. It is glibly assumed too that competent judges should always be in agreement on the ranking of fish in a competitive class. One can easily visualise situations in which this would not necessarily occur. When fish are scored on a numerical points system it is possible that two or more fish may tie; then it is largely a matter of the personal and subjective taste of the judge how the final ranking is achieved. In judging perhaps more satisfactory results are achieved when two or more people are involved in making decisions even if as a result proceedings are rather protracted. In Koi competitions in Japan a sizeable panel of judges may be involved; in Britain Koi are judged often by a small group which includes both established and trainee judges. The drawback of this system is that it is often the judge with the loudest voice or most domineering personality who prevails. In goldfish competitions usually only a single judge is involved.

It makes good sense for the breeder to take careful account of the current show standards in defining breeding policy and establishing selection criteria, especially for quality fish. These selection criteria should be applied in the first instance to the brood stock and subsequently in the progeny. Care must be taken to ensure in the selection of brood stock in the initial and subsequent stages that males and females should be chosen to complement each other rather than in having the highest levels of similarity in appearance. Hence brood stock are not necessarily those which themselves are prize winners at shows but those which produce progeny of high quality. Progeny testing is essential, as experience of many breeders has shown that seemingly indifferent or unremarkable parents may produce outstanding progeny while outstanding champion fish may produce very disappointing offspring. A brief consideration of what may be happening genetically is important here.

Desirable features are not always easily fixed genetically; it may, in fact, be impossible to do so. Some characteristics may be highly heritable and easily fixed, others may be strongly affected by the environment in early development stages and show less consistency of expression. Some may be determined by a hybrid or heterozygous constitution and thus unfixable. As is discussed in more detail elsewhere, nature and nurture are not alternatives and both must be taken into consideration. What breeders may fail to appreciate when spawning champion quality fish is that the combined genetic potential is not necessarily realised in the first generation and breeding should be continued into further generations. By doing so there is a very good chance that promising genetic recombinants may arise and be available for selection. Mendel had to take his crosses to the third generation and beyond to make his points. In goldfish we are dealing with a much more complicated genetic situation and we must be at least as patient as Mendel. We must use high-quality genetic material and raise the progeny under optimal environmental conditions to maximise production of progeny with the most appropriate expression of the desired characters.

The features we will consider are in two classes which can conveniently be labelled 'general' and 'special'. The 'general' characters are those which we would like to see embodied in our fish regardless of variety. The 'special' characters are those which are expressed in some (perhaps only one) varieties and not in others. To consider this question rationally we can follow evolutionary logic and the development of the goldfish from a progenitor through various intermediate forms to the most highly developed.

The Common Goldfish standard

The Common Goldfish (see Plates 1 and 2, wild type and coloured) differs from the wild ancestral type only in coloration. However, coloration is not the only consideration which concerns the breeder because the wild-type phenotype is not completely uniform. Perhaps the most variable features are the body contours, the dorsal and ventral, especially the latter. The desirable contour form is that of a smooth curve from snout to the caudal peduncle. The breeder usually has little problem with the dorsal contour, which in most cases is a satisfactorily smooth curve; the ventral contour is a very different matter and the desirable smooth curve from nose to tail is less common than a rather angular contour brought about by a lack of curvature between the pectoral and pelvic fins and the pelvics and the anal fin. Another important point is that the dorsal and ventral curves should be of approximately equal amplitude. The lateral contours (viewed from above) should be approximately equal, making due allowance for an asymmetry produced in mature females by ovary development. It is necessary to check contours both from the side and from above in selection and judging. In the West, the prime viewing point is from the side and there may be a temptation to pay less attention to inspection from above. This is the prime viewing position in the Orient and inspection from this viewpoint can be very informative indeed, especially when it comes to the assessment of deportment and an holistic appreciation of the individual fish. Fish which appear normal from the side can show surprising deformities in the vertebral column when viewed from above. It is unfortunate that at many shows tanks are often placed too high for the judges to view fish other than from the side in safety unless they manhandle the tanks themselves.

In the case of the common goldfish, it is of course axiomatic that development of all other parts of the body is complete and normal. In the GSGB standard an acceptable range of body depth of 3/7 to 3/8 of the length is set down. The width of spread of the caudal fin is not to be greater than one and a quarter times the body length and the length of caudal fin lobes is not to be greater than one-third of the body length. In both cases the dimensions set out in the standard drawing are less, i.e. the spread of the caudal is approximately equal to body depth while the length of the lobes of the fin is closer to one-quarter than to one-third of the body length. Such guidelines are useful in quantifying such features where this is necessary but the experienced breeder and judge will gauge them by eye at a glance and usually will not need to resort to actual measurements unless confronted with the unusual.

Guidelines for colour vary widely from society to society, perhaps the most comprehensive range is to be found in the *GSGB Standards* (fourth edition 1987). For self-coloured metallic fish red, orange, yellow, blue, brown and black are included while acceptable colours in variegated fish are red, orange, yellow, silver and blue. It is worthy of note that silver is acceptable in variegated but not in self-coloured fish while brown and black are not approved in variegated fish. The reason for this is that black (and brown) in variegated fish has been considered a transitional colour which could be expected to disappear. Recent experience has suggested (in other varieties at least) that this is not necessarily so. Individuals which undergo a late colour change may retain some black coloration for a year or more. A brief comment on self-coloured silver fish is appropriate; these generally are regarded as totally lacking in interest and not worth entering in a show. In a pond, however, they do provide a pleasant colour contrast.

In sharp contrast to the situation in Britain, the Goldfish Society of America (GFSA) recognises only two colour phases, orange-red and red and white. The vast majority of common goldfish entered in shows are of these types and it is only comparatively recently that true yellow goldfish have been introduced there. It would be a comparatively easy matter for breeders to produce blue, chocolate (brown) and black metallic fish if there were sufficient incentive to do so, in addition to interesting possibilities in variegated variants. It is unlikely that more than two colours could be combined successfully but the goldfish continually confounds the prophets and who knows what an inspired selection programme might achieve!

It is very interesting to note that the GFSA has attempted one of the most extensive listings and definitions of colour in metallic goldfish. These are summarised below.

1 Ancient Bronze: brown to black with metallic shine.
2 Black: coal black with no metallic shine.
3 Blue Scale: grey blue to sky blue.
4 Chocolate: very dark brown.
5 Copper: pale brown.
6 Gold: orange.
7 Green: olive green.
8 Mahogany: reddish brown.
9 Red: scarlet to oxblood.
10 White: silvery white.
11 Wild: light to dark greenish brown.

This listing is of considerable interest and is worthy of some comment. A notable absence is of yellow coloration but otherwise the range is quite comprehensive. One difficulty with the classification is that colour development is subject to the influence of the environment. In an open pond environment under appropriate light régimes fish might develop a high intensity of pigmentation, in subdued lighting on the other hand colours would be much paler. Thus an individual fish might be a deep red or intense black in the summer and if moved indoors and exposed to low intensity light could become orange or dark brown. These effects are worthy of extended study to determine how heritable shades of colour are. For example are copper, mahogany and chocolate genetically different or is the difference in shade brought about by differences in the quantity and quality of light received? A very similar situation is found with orange and red where a light induced intensification is a matter of common observation; however, the intense ox-blood colour is thought by some breeders to be due to a genetical difference rather than an environmental effect. It would be equally appropriate to subject the wild-type and 'green' phenotypes to a similar study. It is common practice for exhibitors prior to shows to keep their prized specimens in a colour enhancing environment. This can of course be manipulated to produce the optimum effect. There is another situation in which intensity of coloration can be enhanced: in variegated red and white ('Sarassa') fish the intensity of red pigmentation developed may be greater than in red self individuals. It is almost as though both types produced a similar quantity of red pigment which was spread uniformly in self-coloured fish but concentrated locally in the pigmented areas of variegated individuals. This point is worthy of further investigation as it is commonly believed that substantial intensification occurs.

The consideration of colour in the Common Goldfish has general implications for other varieties of goldfish. While there is an undoubted predominance of the self-coloured red or orange type, there are excellent yellow strains available and it would be possible if anyone wished to undertake the effort to produce black, brown, blue and perhaps stable bicolours in addition to the commonplace red and white Sarassa type. Stable black and red and also black and white combinations could prove to be popular, as could silver fish with contrasting red fins.

The Comet standard (Plate 6)

The ideal Comet goldfish bears a very appropriate name suggesting grace, elegance and movement. Unfortunately it is an ideal which is only infrequently achieved. The body should be streamlined and slender, depth being only 25–30% of the length, the very deeply forked caudal fins should have lobes approximately equal to the body in length held straight and well spread in a V-shaped configuration without sagging. Acceptable Comets these days are rarely if ever seen in commercial offerings by retailers. Those offered as Comets usually have bodies like the Common Goldfish with caudals often only half the length of the body. The original Comet tended to be self-coloured red or orange but more recently red and white or Sarassa Comets have come to the fore. There is an interesting parallel here with Koi where the progenitors of the highly valued and greatly appreciated Kohaku were called Sarassas. In Koi the elegance of a red and white variant, the Tancho Kohaku, is reproduced in the Comet Goldfish. The all-silver or white body is wonderfully set off by a strong oval red patch on the head, resembling in colour and position the similar patch on the head of the Japanese crane. The origin of the Comet has been discussed and as a recognised variety it undoubtedly has an American provenance. One is tempted to speculate whether the Tetsugyo, a Japanese uncoloured Comet-like fish, might not have been involved. A cross with a red-metallic goldfish, recombination and segregation could have produced the Comet in the ponds of the Fish Commission in Washington.

The Comet has the reputation of being a very agile, fast-moving fish, very well suited to life in the pond, less so to the small aquarium. These characteristics have been related by some writers to the size and development of the caudal fin. This belief seems to be at odds with the principles of mechanics. If one looks at the fastest moving fish in nature they have streamlined muscular bodies but not large caudal fins. I venture to suggest that the Comet is a fast-moving fish in spite of and not because of its well-developed caudal fin, which in all probability would produce more drag than propulsive power. To emphasise this point, one has only to consider relative mobilities in twin-tailed fish between short and long-finned individuals. The athleticism of the Comet is probably due to a slender, streamlined but muscular body form.

The Shubunkin standard (Figs 4.3 and 5.1; Plates 7 and 8)

The Shubunkin is the goldfish variety, perhaps above all others, whose history, origins and antecedents are the best known. From its Japanese origin in the late nineteenth century its progress has been monitored scientifically by many, most notably by Matsui. Curiously, it is not a particularly popular fish in the land of its origin, but in the West it is one of the most popular varieties. There is some difference of approach in the judging of this variety on the two sides of the Atlantic. The Goldfish Society of America's Guidelines (1996) consider the Shubunkin to constitute a single variety (or breed as they term it) with three distinct sub-types – the American, the London and the Bristol. Differences in finnage are allowed for in the points schedule. All Shubunkins have the common features of calico colouring and pattern and the body form which is similar to that of the Common Goldfish. In considering variation in finnage type it is appropriate to bear in mind the form of the Japanese original. In this the caudal fin was typically about half the length of the body or slightly more, its lobes were narrow and pointed, pectorals pelvic and anal fins were long and pointed while the dorsal fin height approached in height the depth of the body.

American breeders have essentially taken the Japanese original as their starting point and selected for greater length of fin, so that ideally the dorsal fin height exceeds the depth of body while length of the caudal fin lobes equals or exceeds that of the body. Other fins should be long in proportion. In summary, it can be said that the ideal American Shubunkin is virtually a Calico Comet. Retailers often offer them as Comet Shubunkins; the latter are usually little if at all different from the Japanese original with caudals only approximately half the length of the body.

Whereas the ideal American Shubunkin can be considered as the calico counterpart of the Comet, the London is regarded as a calico version of the Common Goldfish. In British goldfish society standards it is effectively treated as such, as in the old FBAS (Federation of British Aquatic Societies) standards, in the various editions of the GSGB standards, the new

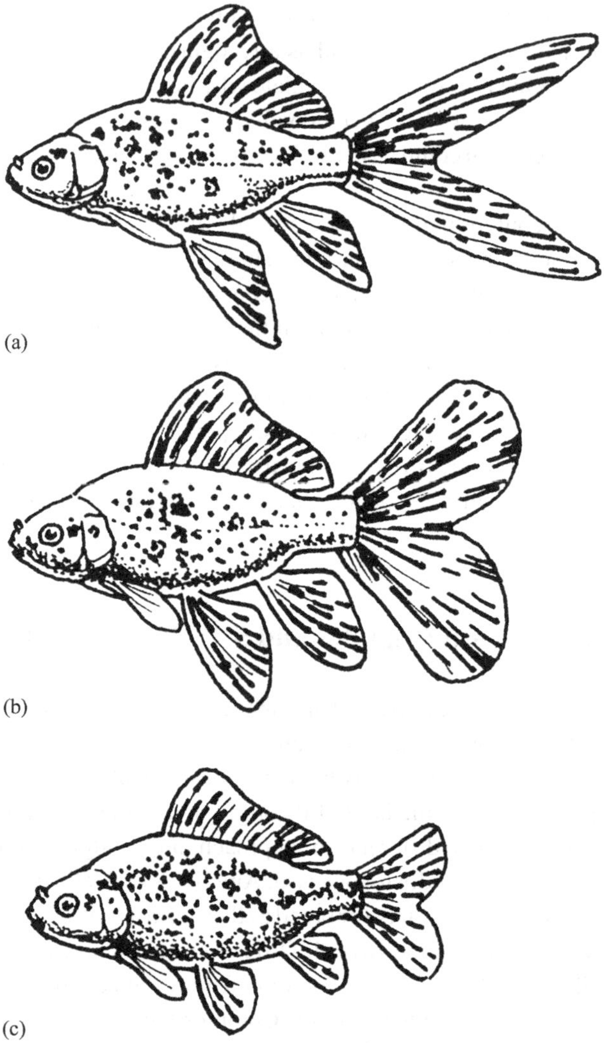

(a)

(b)

(c)

Fig. 5.1 The three styles of the Shubunkin: (a) Japanese/American; (b) Bristol; and (c) London.

Nationwide Standards and in the various society standards which the latter has superseded. At this point we should remind ourselves that the relationship between the London Shubunkin and the Common Goldfish is an indirect one. It did not originate from it directly, by mutation say, but from a probable cross, perhaps in a pond, between imported Japanese Shubunkin stock and an established Common Goldfish population in the south of England.

While the change which produced the London Shubunkin might be considered regressive, involving a reversion to the basic goldfish form, that which produced the Bristol Shubunkin between the wars and which has continued since has been unequivocally progressive. It has produced what has become arguably the most popular variety among goldfish fanciers in the British Isles. It has developed in the country over the best part of the twentieth century and its continued progressive evolution has been sustained by the great appeal it has had for enthusiasts in the country. It is also greatly admired in North America and Europe but for

Plate 1 Wild goldfish.

Plate 2 Common Goldfish.

Plate 3 Wakin.

Plate 4 Jikin.

Plate 5 Tosakin.

Plate 6 Comet.

Plate 7 Bristol Shubunkin.

Plate 8 London Shubunkin.

Plate 9 Sarassa Ryukin.

Plate 10 Calico Veiltail.

Plate 11 Calico Pearlscale.

Plate 12 Black Butterfly Tail.

Plate 13 Red and White Butterfly Tail.

Plate 14 Calico Butterfly Tail.

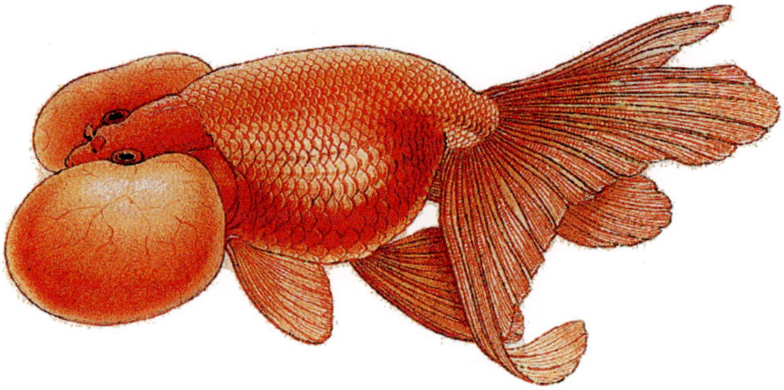

Plate 15 Bubble-eye.

Plate 16 Chocolate Oranda.

Plate 17 Red Oranda.

reasons which are not entirely clear it has never been produced commercially on a large scale; perhaps environmental adaptability is a limiting factor.

The most obvious and characteristic feature of the Bristol Shubunkin (apart from its colour) is the uniquely beautiful caudal fin. In current stocks this shows dorsal and ventral lobes which have an almost circular profile, the trailing edges are certainly semicircular. This feature is combined with increased height of the dorsal, and increased length of the other fins, which have broad-rounded profiles in contrast to the narrow, pointed conformation in the American or Japanese types. An interesting point to note in connection with fin length in the Bristol Shubunkin is that the standards specify that the caudal should not be overly long. In the American type caudal lobes equal to or greater than the body length are desirable, the only limitation effectively is the length which can be supported without sagging or drooping. The same consideration applies to the dorsal fin where the leading spine has to be long enough and strong enough to support the fin without its leading edge curling backwards or to the side. The strength of the supporting spines and rays in calico fish seems to be less than in metallics, it is therefore perhaps unrealistic to expect that comparable fin lengths will be achievable in calico as in metallic fish.

A problem can arise in Bristols when strong selection is practised for fin length in young fish. Fish which show precocious fin development tend with age to become overfinned. Instead of progressive improvement up to an age of four years and beyond such fish may show collapse of finnage, notably sagging dorsals and caudals. The conformation of the Bristol caudal leads to a degree of scissor-like action when the fish moves, while a certain amount is tolerable, excessive scissoring is undesirable and detracts greatly from the quality of deportment in fish prone to it.

The question of colour and pattern is a general issue not only in all types of Shubunkin but also in the generality of calico goldfish. Coloration consists of a basic blue ground which may not be complete, upon which are superimposed substantial areas of bright red or vermilion, dense black spotting and often some areas of silver-white. The scales are predominantly transparent, lacking the reflective guanine of metallic scales; there may be some isolated reflective scales but the fewer the better. Fins should be streaked liberally with black. Guanine is, however, deposited below the scales in a continuous layer producing the characteristic mother-of-pearl or nacreous sheen. Complex mixtures of colours may be present in some fish in addition to those mentioned: violet, yellow, orange, brown and even green. The pattern is not fixable genetically, unlike the body conformation which can breed more or less true. The most desirable expression of the transparent gene is found in the heterozygote (Tt) and as a result Shubunkins do not breed true.

The calico phenotype does not depend only on the transparent gene in the heterozygous state but also on the absence of the gene which produces demelanisation in the Common Goldfish. In the presence of the demelanising gene no blue, black or brown colour will persist; if demelanisation is delayed then these colours though initially present may fade. Transparent scale and demelanising (*Dp/dp*) mutations together can result in the production of similar colour mixtures and patterns such as the Sarassa, in metallic and transparent scaled versions. It is interesting to note that in the immediate postwar period, certainly in the London area, the true calico pattern and coloration was rather uncommon in Shubunkins. It took sustained selection over a number of years to restore in particular the desirable blue background colour.

Blue colour in Shubunkins is particularly valued by American producers and strains have been marketed in which the red colour has been virtually eliminated by strong selection. In Britain such fish were sold as Cambridge Blue Shubunkins. It is rather ironic that the name Shubunkin in Japanese implies the presence of a strong vermilion element in its coloration! A novel breeding objective for the Shubunkin, the production of a more-or-less true-breeding form, which only a few years ago seemed to be but a pipe-dream, looks as though it may yet come to pass due to the discovery of highly coloured homozygotes (*TT*) with the depth and colour quality of the Shubunkin. Until these came to light, fish homozygous for the transparent gene tended to be devoid of colour ('pinkies') and not of great interest. This topic will be considered at greater length later.

The ideal Shubunkin aesthetically speaking would show a predominance of blue and violet in the background, overlain with strong red, orange and yellow, with well-distributed spots and blotches of intense black or rich brown. The important features would be a pleasingly balanced pattern of colours and distribution of the dark spots. This would be combined with the appropriate single-tail body conformation (like the Common Goldfish) combined with finnage appropriate to the Japanese–American, London or Bristol types.

The Fantail standard (Fig. 5.2)

The Fantail Standard provides a point of reference in a sea of variation which has become more extensive in the course of the twentieth century. The Fantail of the standards is in fact a western fish, the nearest approach to which in Far Eastern stocks is the Watonai, a hybrid between the primitive twin-tail, the Wakin, and the more highly developed Ryukin.

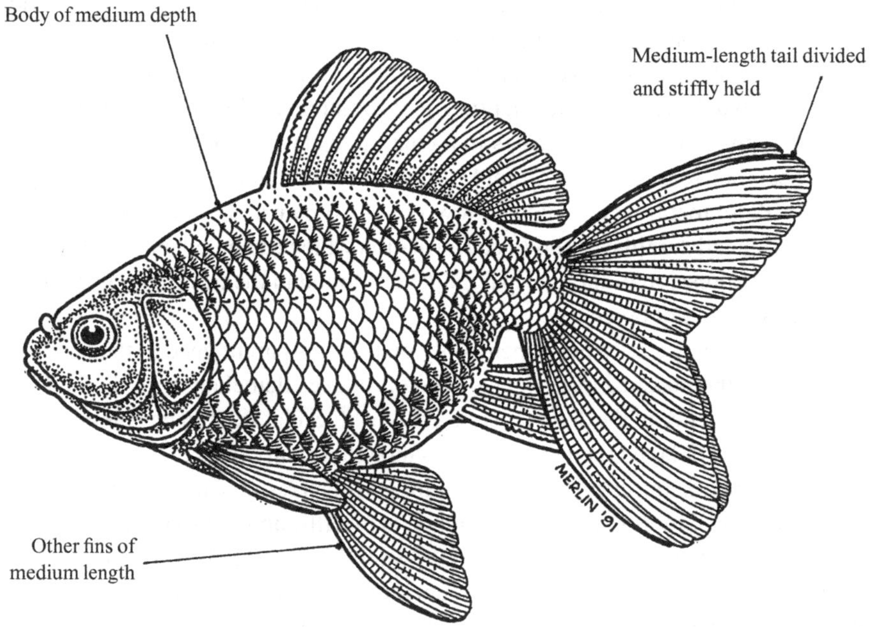

Fig. 5.2 The salient features of the Fantail.

As we have considered, the Wakin is fundamentally a twin-tailed Common Goldfish, with increasing age tending to develop greater depth of body. Hervey and Hems suggested that the twin-tail mutation tends to promote body depth and set in train the developmental processes which culminated in the extreme body depth we see in the modern Ryukin. The development of both body depth and length of finnage is controlled by multifactorial or polygenic systems which means that when these features are studied and measured we find that variation is continuous and we do not have segregation into distinct and discrete classes. If we want to categorise such variation we have to set rather arbitrary limits to what we consider to be the acceptable part of a much wider range.

The modern Fantail has a body depth (according to standards) of 60–70% of its length; the length of caudal is 40–60% of body length. The longitudinal axis of the fish is horizontal and the tail is held erect without any trace of downward curvature of the caudal peduncle. The caudal fin is bifurcated and the body tends to be somewhat globular in form. This standard is an advance on the Wakin with respect to body form, which is less compressed laterally and appreciably shorter in relation to overall length and with longer finnage. It is interesting to note that in Innes' book there is a photograph of commercial Fantails which are intermediate in body length between the Wakin and the modern Fantail. This serves to illustrate that sustained selection over decades brings about significant change by the accumulation of often imperceptible small changes. There have been few if any attempts to produce show standards for the Wakin but there should be few problems in judging such entries in the AOV (Any Other Variety) classes at shows. One can regard the Wakin as a twin-tailed Common Goldfish. While most Wakins show typical Common Goldfish type metallic coloration, there are reports from Japan of strains with very intense coloration. These are bicoloured red and white with a particularly intense deep shade of red which not uncommonly develops in Sarassa patterned fish. The Wakin will always be of interest as the progenitor of the whole range of fancy goldfish we appreciate today.

The modern Fantail is a popular fish which is hardy enough to flourish in a pond environment. Both metallic and calico forms find favour with breeders. In recent years there has been development of particularly deep-bodied Red Metallic Fantails. These fish have relatively short fins in relation to body depth, a combination which some might regard as less pleasing than the combination of less body depth and somewhat longer fins. The best Calico Fantails which have been entered in shows of recent years have included some excellent specimens, with pleasing but not extreme body form and finnage and glorious colour: brilliant reds, bright blues and excellent patterning of dark spots and streaks. The Fantail has a great advantage in not being a particularly highly evolved form of goldfish, in that it is less susceptible to problems brought about by change in body form such as digestive and swim-bladder problems. It is certainly a delightful variety and perhaps one in which a good balance is achieved between the extent of its development as a fancy fish and its general hardiness and ability to hold its own.

As the basic twin-tail fish the Fantail epitomises certain features which are common to the group as a whole with regard to the caudal fins, which should be duplicated completely and their dorsal margins separate for approximately 90% of their length. Some exceptions to this general rule will be considered in the appropriate place. Duplication of the anal fins is considered to be of great importance in the West but much less so in the Orient.

The Jikin standard (see Plate 4)

The Jikin is remarkable for two features: its splayed twin-tail and the striking colour pattern. The form of the caudal fins is unique in that they are held virtually at right angles to the caudal peduncle and are dumb-bell shaped. The body is laterally compressed with moderate depth but not foreshortened and without any tendency to globosity. Colour of the Jikin can be very striking indeed with intense deep red fins (including the caudals), red lips and perhaps some red along the ventral contour. Perfect Jikins are difficult to produce; the heritability of the caudal fin conformation is not high, neither is that of the colour pattern. As noted previously, attempts have been made at times to improve on nature by cosmetic surgery, which is to be deplored. Perhaps over the past two or three decades breeders have become rather disillusioned at the lack of success and their numbers have declined. The Jikin colour pattern arises sporadically in other goldfish varieties and can be very striking; it is remarkable that it has become established as an essential feature only in this variety.

It is unfortunate that the Jikin is spite of considerable efforts over recent decades has proven such a difficult subject. It is to be hoped that some means can be found to improve the efficiency of its production. It would be a shame indeed if we were to lose such an unusual goldfish.

Mention should be made of the attempt to produce a calico Jikin – the Edo-Jikin. The examples produced were striking and beautiful fish generally similar to the Jikin but the tail region did not reproduce exactly the caudal conformation of the true Jikin and at best seemed to be more of a Calico Wakin. In terms of its form, a twin-tailed fish with an elongate body, the Jikin clearly has a close affinity with the Wakin. The only Jikin standard I have encountered was produced by the FBAS in 1988.

The Fringetail standard (Fig. 5.3)

The Goldfish Society of America at one time produced a Fringetail standard as well as one for the Ryukin. The two names have been regarded as virtually synonymous since the Fringetail of a century ago was clearly the progenitor not only of the modern Ryukin but also of the very different modern Veiltail which was called 'Japanese Fringetail' in Innes' book. This situation highlights some problems in naming varieties: both names and the fish themselves can change with time. The same fish may be given different names in different places, or the same name may be applied to two quite different fish. A decision may be taken to give a totally new name which may solve the problem if it is accepted, but adds to the confusion if it is not. I propose a redefinition of Fringetail to cover fish which are not better described as Veiltails or the more highly evolved modern Ryukin. These would be characterised as having lesser development of the hump than in the Ryukin and deeply bifurcated caudals with long and narrow lobes. The term 'Ribbon-tail' is often applied to examples with particularly long, narrow caudal fin lobes which can create a very attractive effect as they stream out behind the fish as it moves forward. Depth of body is not extreme but should exceed half the length.

Fringetails may have metallic or transparent scales. As is the case with other varieties, red, orange and red and white are the most common colour variants but good strongly coloured calicoes are also to be seen. This can be a good potential source of stock to improve colour in fancy twin-tailed varieties which have shown deterioration in quality and intensity of colour.

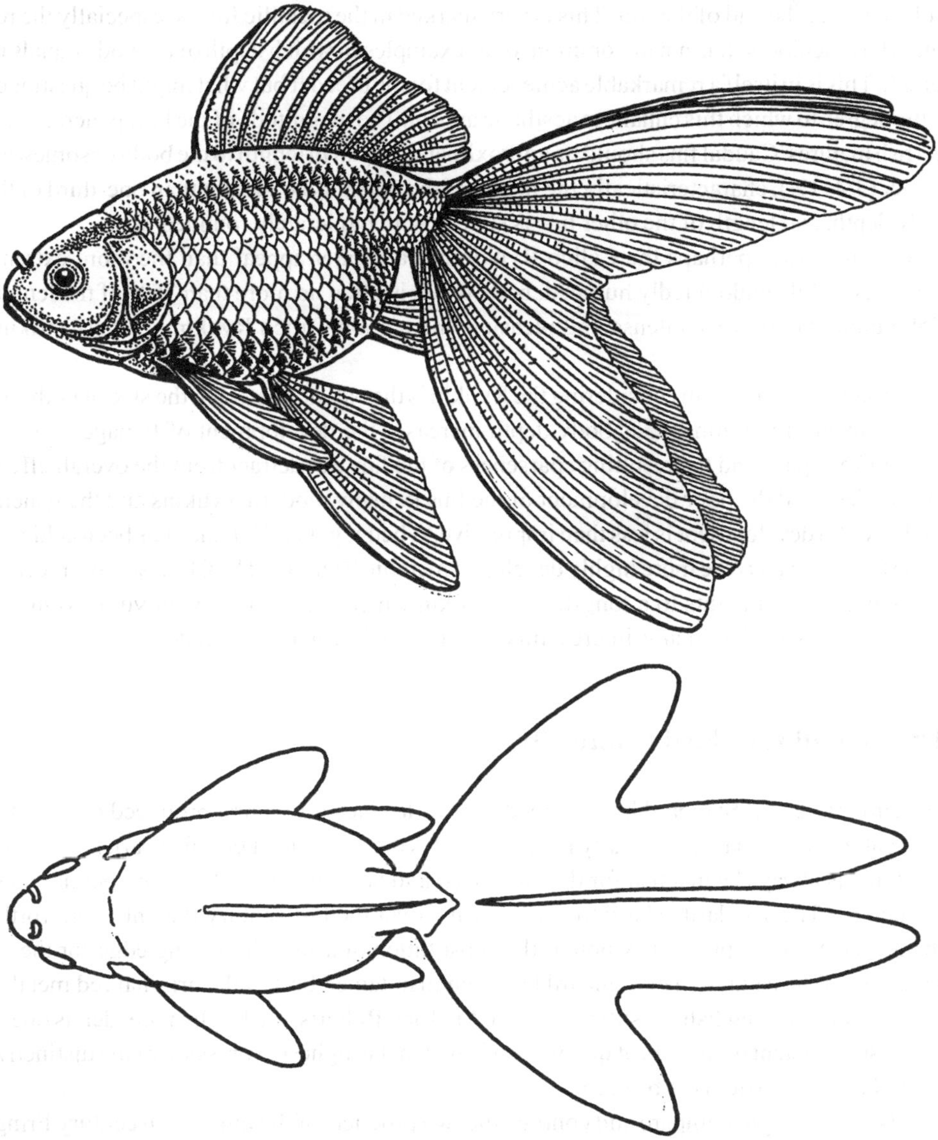

Fig. 5.3 Fringetail.

The Fringetail, by virtue of the fact that it has given rise directly to other varieties such as the Oranda and Tosakin in addition to its own merits, is deserving of our attention. It would be a great pity if it were to disappear as a consequence of neglect.

The Ryukin standard (Plate 9)

Ryukin Standards have been produced in America but none have been developed in Britain. The most obvious distinctive feature is the hump-back which has been developed enormously in Japan since the end of the war. This is very marked in the metallic forms, especially the red and white bicolours. It is not uncommon to see examples in which depth of the body equals its length. This is in itself a remarkable achievement for the breeder but what might be questioned is the degree to which this enhances aesthetic appeal. The magnitude of the hump increases as the fish matures. Caudal fins should be approximately equal in length to the body or somewhat longer (3/4–1½), characteristically the height of the dorsal is no greater than one-third of the body depth, and length of the other fins is similar with moderately rounded tips.

To some tastes perhaps the Calico Ryukin has a greater appeal than the more extreme metallics. While undoubtedly humped, body depth is often no more than 75% of the length. Coloration may be very intense and pleasing and a good Calico Ryukin is a very striking fish.

A point which arises in considering the Ryukin is the balance between the size of body and development of the finnage; as body depth increases the development of finnage does not seem to keep pace and the resulting sparseness of the fins can detract from the overall effect. Matsui deplored the poor development of the fins in many modern Ryukins and the general quality of fin development is less than impressive. A better general balance has been achieved in the Calico *vis à vis* the more highly developed metallic Ryukins. The Chinese have recently developed what amounts to a long-finned Ryukin which is a form of Wen-yu; it would be interesting to see what, if any, interest this might be to the Japanese breeders.

The Veiltail standard (Plate 10)

The current Veiltail Standard has changed very little since it was first produced early in the last century and to many, especially in the West, it is the pinnacle of goldfish variety development. It has always been a fish for the enthusiast and has not lent itself to commercial mass production. The standard is hard to maintain and over the last century the only noteworthy change has been (as previously noted) the substitution of a straight trailing edge for the indentation of the caudals. The standard has been maintained best in the original red metallic form and outstanding fish are still to be seen: Mr Tony Roberts, the English breeder, is one of the most consistent producers of quality fish in which the highest expression of the distinctive Veiltail characteristics is to be seen.

It is interesting to compare and contrast the two products of the nineteenth century Fringetail, namely the Ryukin, and Veiltail. In the Ryukin body development in its extreme form has occurred at the expense of the finnage. The Veiltail shows development of body depth to a more modest extent of approximately two-thirds to three-quarters of body length. Dorsal and ventral contours are as smooth as possible with the least possible suggestion of a hump. It is, however, in fin development that the Veiltail is supreme; something has already been said of caudal fin development, which should be very broad indeed in order to produce the much-admired Veiltail effect in which the caudals hang in elegant folds when the fish is stationary and spread like a banner when it moves slowly forward. No less remarkable is the dorsal fin, which should be high, equalling body depth with a convex dorsal curve; the trailing edge should be more or less horizontal giving a sail-like effect. All the other fins should be long and hang gracefully. This mass of finnage produces tremendous drag and reduces mobility, which make a mature Veiltail unsuited to any but a very protected environment. Breeders who attempt to exploit the pond environment to enable brood stock to recuperate, all too often find their Veiltails fall easy prey to a very wide range of predators from cats to crows. Vulnerability to predation and physical damage is a very real hazard in the more highly developed varieties.

While the red-metallic Veiltail is, in the view of the writer, unsurpassed as the most elegant and refined goldfish variety, it is interesting to note the further development of the Veiltail which has been attempted. The most noteworthy and one to which a great deal of attention has been given is the Calico Veiltail, which has provided the greatest challenge. Considerable and devoted effort has gone into its development and maintenance. It must be said while the metallic Veiltail is holding its own, the Calico Veiltail is having a rougher passage. Veiltails between the wars were widely distributed in Europe and of course in North America. They were virtually extinct in mainland Europe at the end of World War II and survived only in Britain. In North America after the war, enthusiasm for the goldfish faded with the rise of the cult of the Koi and it seems that the genetic base of the famous Philadelphia Veiltail became very restricted and the pre-war standard was not maintained. The situation has been in part retrieved by importation of British Veiltails (themselves derived from the Philadelphia strain). The most noticeable difference is that depth of body is less than it was in the early days and the sheer majesty of the best Veiltails has not been regained. The new style slim-bodied fish are certainly not without appeal but they are a rather different fish from the original.

The combination of the Veiltail conformation and the transparent scale mutation has had consequences which have only recently been fully appreciated. Americans are sometimes surprised at the relatively short caudals of British Veiltails. The reason for this apparently is that there is a side (or pleiotropic) effect of this mutation which results in rather more delicate and less robust fins and fin-rays, which are also more prone to bacterial infection. This has been discussed already in relation to the Shubunkins; the finnage of the original Japanese Shubunkin can support itself well and the longer-finned derived American type also because the lobes are narrow. In the heavier Bristol tail there seems to be a definite upper limit to size. Similar considerations apply to Calico Veiltails; the length and breadth of caudal fins which can be supported in metallics may not be practicable in calicoes. The early Veiltails did not develop their outstanding features as young fish but only as they matured. The modern tendency to select for precocious development is to be resisted; such fish are unlikely to maintain their attributes for long and their peak of optimal development may also be short-lived. Short finnage (within reason) in Calico Veiltails could in fact favour maintenance of quality in the longer term since in adults fin growth persists at a more rapid rate than that of the body.

The high dorsal of the Veiltail demands a very robust development of the supporting spiny ray. If this lacks adequate strength then it may bend sideways or back and the sail-like effect would be lost. It is very important that development of height and length of fins does not go beyond what can be supported in the mature fish.

Colour development in metallic Veiltails does not present any problem but it has become noticeable of recent years that the strength of colour in calicoes is not as great as it has been in the past. It is possible that this may be the result of crosses with metallics followed by back-crossing to calicoes. The effect of the depigmenting genes would be to reduce the intensity of the desirable blue coloration. Selection for the presence and strength of blue colour would reduce the frequency of depigmenting genes but there is the possibility that if any survive in the population they could produce depigmentation. It is a useful check to retain a sample of metallic segregants from a calico spawning and to see how many of these depigment. I have had metallic progeny from both London Shubunkins and Calico Veiltail spawnings which depigmented. Incidentally, the female parent London Shubunkin used lost its blue colour as it advanced in years.

An interesting point emerges regarding the carriage of the caudal fins if one compares a Veiltail with a Nymph from the same stock. Nymphs these days are totally ignored but it is interesting to note in most, if not all, the drawings made of the type in older books that the carriage of the single caudal is virtually horizontal, compared with the pendent caudals of the Veiltail. In the Fringetail and Ryukins carriage is closer to the horizontal than in the Veiltail; it is not entirely clear what produces this effect – perhaps it is a correlated growth response of the caudal peduncle and the fins it supports.

The Tosakin standard (see Plate 5)

The Federation of British Aquatic Societies published in 1988 a standard for the Tosakin, one of the very few which have been produced and which in the light of the variety's fortunes was produced in a spirit of optimism. The Tosakin is unquestionably an exceedingly beautiful fish seen often in photographs but rarely in the flesh. Many of those published are truly exquisite. Unfortunately the fish is a very difficult subject indeed; it is not easy to keep and a problem to breed, while individuals showing the full development of the very distinctive tail are very rare.

In basic conformation the Tosakin is close to the Fringetail; body depth should be in excess of 50% of the body length without a pronounced humpback. Among fancy goldfish it is unusual in that possession of a web-tail is a *sine qua non*. The joined dorsal lobes extend backwards along the upper surface, the trailing edge of the fused caudals then curves forward, the point of greatest indentation of the caudals is approximately level with the caudal peduncle, the upper portion of the ventral lobes continues the forward curve until the lower portion is sharply recurved backwards towards the free end. The form of this curve is dependent on the anatomical structure of the caudal peduncle; any imperfections and asymmetry in its development would profoundly affect the carriage of the caudals.

It is tempting to link this variety with the Jikin as being particularly difficult; of the two the Tosakin is the more exacting. It would probably be useful to examine and analyse carefully the conditions under which this variety was developed in Kochi City (formerly Tosa) and under which it has been produced in recent years. It is known that in China production of some specific variants is only practically possible in very restricted geographical areas, the reasons for which are not known but presumed to be environmental. In the West we experience this problem in an even more severe form. To overcome it would require an extensive (and probably expensive!) study of the development of characteristic features under varying environmental conditions.

Due no doubt to the difficulty of reproducing the basic type, there has been very little attempt to ring the changes regarding colour. Tosakins are self-coloured red or red and white on the body sometimes with contrasting fin colour, for example, black. An interesting variation is that the trailing edge of the caudals may be scalloped instead of having the more usual straight edge.

The Pearlscale standard (Fig. 5.4; Plate 11)

Apart from the distinctive form of scale after which the variety is named, its most characteristic feature is the extremely globose body, which looks almost pathological, as though the fish was dropsical. Otherwise the body proportions are similar to those of the Fantail. The scales differ from normal in being domed, having the 'pearl' at the centre of the exposed portion. The uniformity and strength of this development are important features together with the absence of normal scales. Both metallic and calico types occur and good calico fish can be very impressive indeed. Finnage in Britain tends to be short and Fantail-like but in America examples with rather longer finnage and square-cut caudals are also found.

The pearlscale character has been recombined with features of other varieties most notably to produce the Pearlscale Oranda or Hamanishiki in which the hood takes on an unusual form. This is the form of a large vesicle with a more or less central constriction and is a very distinctive feature. The pearlscale character is one which combines easily with the distinctive features of other varieties and could be an overall enhancement rather than a gilding of the lily.

The Pompon standard (see Fig. 4.18)

The distinctive feature of the Pompon is produced by the proliferation (hypertrophy) of the tissues of the nasal septa to produce what have been called 'narial bouquets' or 'velvet balls'. The established form in the West lacks a dorsal fin but a Japanese Pompon, the Hanafusa, possesses one. In reality these are two basically different fish; the Hanafusa is a Fringetail with narial bouquets, and differs in body proportions, and type of finnage, from the dorsal-less Pompon. The situation is further confused by the fact that the British standards were based on an Eggfish body conformation which differed appreciably from the fish of the variety which were actually available at the time, in which body depth was only one-third to one-half body length and length of caudal fin was over half the body length. The standards set out a breeding ideal rather than a yardstick for judging those fish which were available. The variety itself has not achieved a high level of popularity and an insufficient number of breeders have been attracted to it to bring about any significant change in the direction of the standard prescription.

As a dorsal-less fish, the dorsal contour is ideally smooth with absolutely no vestiges of fins or spines and devoid of lumps and bumps. The pompons should be of a relatively dense, compact texture and as large as possible consistent with this. They may be of the same colour as the body or one which contrasts with it.

It has proven easy to recombine the pompon character with those of other varieties; these recombinants are more successful when there is adequate space for the feature to develop and be displayed.

The Telescope-eye standards (Plates 12–14)

There are regional differences in approach to the Telescope-eye group of varieties. In Britain the GSGB and Nationwide standards recognise two distinct groups, the Moor and Globe-eye while the FBAS recognises only the Moor. The GFSA recognises a more extensive colour range but a narrower range of variation in finnage. The situation has become further confused in recent years by the advent of the Magpies, Pandas and Butterflies which have demonstrated considerable popular appeal.

It is interesting to look back two centuries to the Billardon de Sauvigny scroll where we can see telescope-eyes in abundance, combined with long and short bodies; finnage, however, is invariably short. If we look at the form of the Telescope 100 years ago what we see represented is either a Fantail or a Fringetail with protuberant eyes. The Telescope as a commercial fish has generally been a Fantail, but high-quality long-finned Telescopes have been around for the past 100 years and have formed the basis of the GSGB/Nationwide Globe-eye standard, but with a narrow range of approved colours: only black metallics and calico are acceptable. The other standard is for the Broadtail Moor which by definition is a black metallic telescope-eye, with the body and fin characteristics of the Veiltail.

The most significant telescope-eye fish in the evolution of fancy goldfish generally has been the Calico. This has been the actual source of this colour pattern in all other calico goldfish variants from the Shubunkin to the Veiltail. At the present time it is unfortunately in a state of eclipse, undeservedly so because it can be an extremely attractive fish with pleasing form and strong colour. The reason for this is undoubtedly the extraordinary popularity of the Moor, which many consider to be the Telescope *par excellence*. There is no doubt that a Broadtail Moor in which all characteristics are well-developed and expressed is a supremely elegant fish. The colour should (surprisingly for a metallic fish!) be a dense, solid matt black extending over the whole body and fins. The eyes should be large and well-developed and most effectively globular in form. Otherwise the character of fins and body should be that of the Veiltail, resulting in a truly majestic fish.

It is quite apparent that the amount of variation generated in this variety has raised problems in establishing consensus in setting up standards and determining what those standards should be. There has been a tendency to be rather over-prescriptive with the result that many excellent fish are placed at a great disadvantage because they do not conform to all the minutiae of the standard prescription. This difficulty arises from the extent of variability for many of the characters found in the Telescope. The eye itself varies considerably in size and shape. The size of the eye in the past has been rather smaller in the Telescope than in the Celestial, but in recently developed Chinese types size of the eye is comparable in both. As regards shape of the eye the GSGB standard recognised four types, the truncated conical, spherical, segmented sphere and the ovoid. Perhaps in order that the two recognised standards be kept distinct, the eye type designated for the Globe-eye was the truncated cone while the Broadtail Moor prescription was for the spherical form. The other major distinction is that the Globe-eye tail is forked while that of the Broadtail has a straight edge. Finnage commonly seen in Telescopes could be regarded as approximating to the Fantail, Fringetail or Veiltail in type. The recently developed Butterfly-tail is similar in shape to the Veiltail but the carriage of the caudals is quite different; they are held in a horizontal plane and resemble a butterfly with spread wings when seen from above.

The increased diversity to be seen in the Telescope does, as already noted, create problems for those devising and defining standards. Perhaps the easiest way out of this dilemma is to adopt as broad a standard as possible so that any Telescope fish of quality, ideally, could be eligible for entry. If excessively large entries in the Telescope class were to create problems then the class could be divided on appropriate lines pragmatically. If, for example, half the entries were Moors and the other half Butterflies, each half could constitute a class and be judged separately. Over-prescription should be avoided as we cannot accurately predict changes in fashion or popularity of particular goldfish varieties even in the relatively short term.

The Celestial standard (see Fig. 4.16)

The Celestial's unique characteristic is the upwardly directed telescopic eye, which has led the Chinese to call it the Sky-gazer. This feature is generally combined with a dorsal-less body. There are British and American standards of which the current GFSA (Goldfish Society of America) Guidelines represent very accurately the best type of Celestial that one is likely to see at the present time. In contrast the British Standards (Nationwide and FBAS) project perhaps more of an ideal than an actuality. The two sets of standards differ in two particulars: body length and length of finnage. The British Standards specify body and finnage of the Maruko type with a short body and short fins while the Americans depict a type with a more elongated body and longer fins. There is a possibility that the two types of Celestial both exist in China and that the preponderance of the long-bodied Celestial arose at the time when Hong Kong was the only effective outlet for Chinese goldfish, and Southern Chinese types tended to predominate which had longer bodies. British Standards have probably been based on a type which has now become uncommon. Some British breeders, notably the late Ted Metcalf, made considerable progress in breeding towards the British Standard type.

A notable feature of the best current examples of the Celestial is the beautiful dorsal contour which is not uncommonly seen without any sign of spikes, fins, lumps or bumps, truly a fine sight. The curvature of dorsal and ventral contours tends to be markedly different, greater on the ventral than the dorsal surface. The greatest problem in producing good Celestials is perhaps in the development of the eyes. Their diameter is generally greater than in the Telescopes, the volume of the eyes must be equal, they must both face upwards equally, and size of pupils and iris must also be equal. This end result can create problems when the rotation of the eyes and the migration of the pupils are not perfectly co-ordinated. In some individuals the eyes face laterally and do not rotate.

Most Celestials are metallics with a range of colours; calico Celestials are known but are much less common. The Celestial provides the breeder (especially the British breeder!) with a very considerable challenge. Unfortunately for the variety many do not find it very appealing or interesting and its devotees are a limited if dedicated band of enthusiasts.

The Bubble-eye standard (Plate 15)

A similar divergence in British and American Standards occurs in the Bubble-eye as in the Celestial. The American perspective seems to be based on the current Southern Chinese type while the British has perhaps been based on an earlier and now less common form. It may perhaps, as suggested elsewhere, be more aspirational in nature, a subjective ideal. There is a very close similarity in the body forms of the best Chinese Celestials and Bubble-eyes.

The name of this variety refers to the large infraorbital vesicles which can vary considerably in size. There is a tendency for the development of the bubbles to displace the eyes so that these face upwards as in the Celestial; however, the eyes are not enlarged as in the Celestial. A form of Bubble-eye, the Toadhead, in which the bubbles are not large, is seen occasionally. Smaller bubbles may be more or less spherical in shape but the larger ones do not have a constant shape and wobble as the fish moves. The bubbles are delicate and easily damaged; the damage may heal but the bubbles may not regain the original size.

A wide range of the usual metallic colours are available and some particularly fine Calico Bubble-eyes have been obtained from China. It is a very interesting fish, a great challenge to the breeder, but not to everyone's taste.

The Oranda standards (Fig. 5.4, Plates 16 and 17)

The Oranda, if one accepts Matsui's view, arose as a hooded variant of the Fringetail. It is possible that the original form of the hood was what we know as the goosehead, that is hood development is confined to the cranium. This is a reasonable conclusion to draw because non-

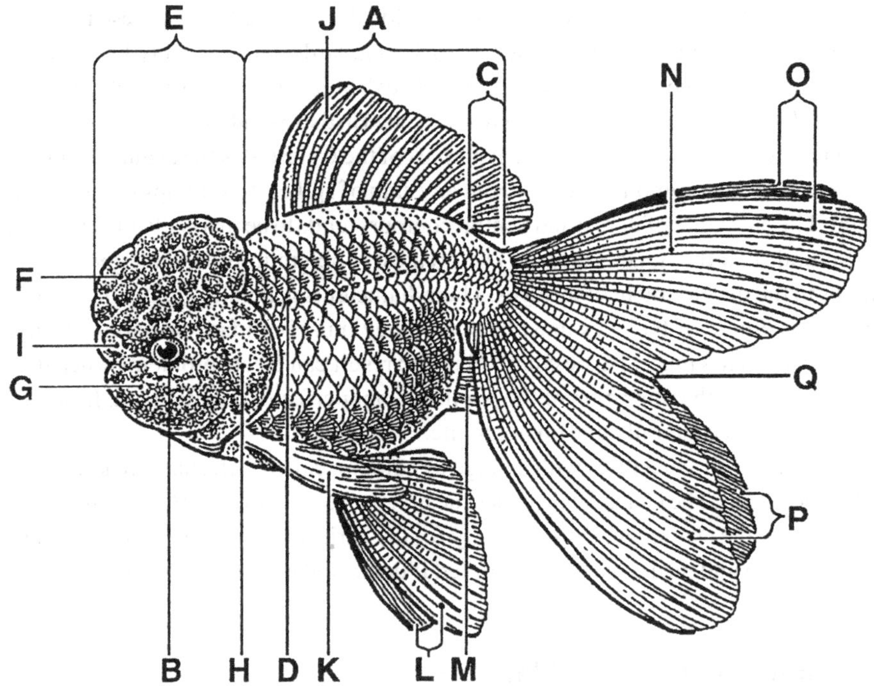

Fig. 5.4 The salient features of the Oranda. A, body; B, eye; C, caudal peduncle; D, lateral line; E, head; F, cranial region (of head); G, suborbital region (of head); H, operculum; I, nasal septum; J, dorsal fin; K, pectoral fins; L, pelvic or ventral fins; M, anal fins; N, caudal fins; O, upper caudal lobes; P, lower caudal lobes; Q, fork.

hooded varieties such as the Veiltail sometimes show a rudimentary hood-like development on the cranium. The two other areas in which hood development can occur are the operculum and below the level of the eyes – the infraorbital area. Development of hoods covering all three areas is commonplace in modern Orandas and some magnificent examples are to be seen. The current popularity of both goosehead and fully hooded Orandas has posed problems in the definition of standards. In the FBAS two Oranda standards are set out: the Oranda with full hood development and the Tancho Oranda which is a Redcap goosehead. The GSGB/ Nationwide Standard sensibly accommodates both in a common standard. This would allow division of a huge class with substantial entries of both types into two separate groups to facilitate judging. American practice is to have a single class for Orandas.

In the past 50 years or so there have been other changes in the Oranda which have produced marked changes in both body form and finnage as well as in the colour range. Pictures of the Oranda 90 years and more ago show a fish with a body depth of half to five-eighths of the body length, somewhat similar to that of the Veiltail. There is a trend for breeders to select for Veiltail-style caudals as well. In a word, following the most recent change in the Nationwide Oranda standard, the Oranda has become essentially a hooded Veiltail. In terms of colour and pattern the Oranda is remarkable in that virtually the entire range of metallic colours are to be seen: red, orange, bicoloured red and white, black, brown (chocolate) and blue; lemon yellow Orandas have not yet been produced in quantity. An interesting feature of the blue Oranda is that hood development tends to be sparse. The Calico Orandas (Azuma Nishiki) which have been produced recently have included some truly spectacular fish; a notable and much photographed example was the truly splendid 'Stoney'.

The Oranda has the capacity to grow to a large size and if quality characters do not deteriorate markedly with age some truly spectacular fish can result. Some recent developments in the Oranda can be noted at this juncture. The combination of the pearlscale feature with those of the Oranda has resulted in the extraordinary hood development of the Hamanishiki in which a large vesicle is developed which has a central constriction which can produce what may appear to be a paired structure (Fig. 5.5). It is possible that this is produced by a pleiotropic effect of the pearlscale mutant in a parallel kind of interaction which produces the extremely globose body of the typical Pearlscale. Another interesting development is the production of hoods whose colour contrasts strongly with that of the body. Red hoods combined with yellowish or rather brassy bodies have been known for some time (especially in Lionheads) but a black hood on a red fish is truly a novel combination. Red hoods contrasting with a white body are of course an established feature of the Redcap or Tancho Oranda, but white hoods contrasting with a red body and red hoods with a black body are novel indeed. Multicoloured hoods, a feature of Calico Orandas, are now appearing in metallic scaled fish.

Variation in the texture of the hood itself is notable in recent Chinese developments, this may be uniformly fine or coarse or it may vary between the three regions of the head, coarse on the cranium for example and fine in the other two regions. It would seem that the capacity for changes to be rung in the Oranda is greatly in excess of what might have been imagined 50 years ago. It remains to be seen if this has been exhausted.

Fig. 5.5 The Hamanishiki.

The Lionhead standards (see Figs 4.22–4.24)

The Lionhead in the broad sense embraces a wide range of distinct forms which have the common characteristics of a short deep body, small fins but no dorsal fin. While the division of the caudals in all other twin-tails (except the Tosakin) should be as complete as possible (say 90% divided) that in Lionheads is often very much less (25% division is specified in the American standard). In the Osaka Ranchu there appears to be no division at all along the dorsal margin of the caudals, that is to say it is a web-tail.

Japanese breeders seem to have had the greatest responsibility for the development of Lionheads as a whole. Setting on one side the question of the hood, the Japanese have developed a remarkable and distinctive body form commonly known as the Ranchu. This is characterised by an extreme curvature of the back and the vertebral column but remarkably, in spite of this, the form of the caudal peduncle is such that the carriage of the caudals is still more or less horizontal. The alternative style (sometimes called the Chinese) has a much gentler curvature of the dorsal contour with a lesser body depth (five-eighths of length) than the Ranchu (three-quarters body length).

The treatment of these two variants in terms of standards has varied. GSGB and Nationwide have essentially adopted the 'Chinese' standard but at shows fish of both styles have been judged together. On the other hand FBAS and GFSA have separate standards for Lionhead and Ranchu. In practice GSGB/Nationwide have made the common standard work and this could be a useful guide for the future. In practical terms in a class with a very large entry, Ranchu and Lionhead sub-classes could be judged separately to accelerate the process if necessary.

The Ranchu is a fish with a dedicated and devoted following in Japan and elsewhere. Its extremely strong divergence from the progenitor type has had some very important implications for the breeder. The absence of the dorsal fin in combination with the egg-like body and the massive development of the skull, especially in terms of its breadth, produces a tendency to a lack of balance in the fish. This is manifested in what is generally known as 'headstanding', to which the Ranchu is particularly prone. This has been counterbalanced by selection for a massive development of the caudal peduncle which helps to maintain a horizontal alignment of the body in movement and at rest.

Hood growth in Lionheads generally shows a range of development. There are indications that the initial hood form was comparable to the goosehead in the Oranda. Curiously and unexpectedly the attempt to produce a Calico Ranchu has resulted in the Edo-Nishiki which has rather poor hood development. This contrasts with the situation in the Oranda where Calico Orandas frequently have magnificent hoods and blue metallics tend to show poor hood development. The Chinese distinguish two types of hood which they call Lionhead and Tigerhead. The texture of the Tigerhead hood is generally fine and uniform; its most important feature is that the hood covers the head uniformly and the cranial, opercular and infraorbital areas are not as distinct as in the Lionhead.

The range of hood colours found in the Lionhead is not as extensive as in the Oranda but handsome Redcaps have been produced and bicoloured hoods (red and white) are to be seen. The Redcap or Tancho feature is one which has featured even more extensively in development of varieties in Koi than in goldfish. There are other features of Koi which also appear from time to time in goldfish especially Japanese Ranchu; one such which is worthy of note is the Kanoko scale in which the central portion of the scale carries a more or less circular spot of red pigment surrounded by silver. Occasionally Ranchu may be seen showing pearl Ginrin scales. These parallel developments in Koi and the goldfish are both intriguing and interesting and are perhaps indicative of areas in which further development of the goldfish might be possible.

Goldfish aesthetics and standards – some conclusions

Standards are useful in ways beyond that of guiding judges at competitive shows. They provide benchmarks for breeders, both commercial and amateur, who wish to improve the quality of their stock in a general way. The commercial breeder who produces stock of show quality can sell at a premium price. We have seen in recent years fish being taken from retailers' tanks and after a very short interval being entered in shows and winning major awards. The value of good standards which embody a consensus view is almost inestimable. The Veiltail standard is a case in point; the salient features were accepted within a couple of decades of its establishment and have not changed substantially since. The agreed standard has been a challenge to maintain and it seems to represent a kind of end-point. To many Western goldfish fanciers the Veiltail is the embodiment of perfection and to seek to 'improve' it beyond the prescriptions of the standards would be to attempt gilding the lily. One might argue that this has been attempted in the convergence of standards for the Moor and Oranda with that of the Veiltail.

The objectives of the founding fathers of the GSGB in setting up their first standards was to establish a set of basic varieties which could not be produced by hybridisation between them and which did not embody too many special characteristics. It was also envisaged that it should not be possible to pass off sub-standard examples of one variety as a different variety. These are very sound basic premises; however, as the number of standard varieties increased then it became very difficult at times to apply these cardinal principles. This difficulty is compounded by the convergence of standards and the production of what have come to be identikit standards. Basically Broadtail Moor and Orandas are Veiltails with additional special characteristics, so much so that a Moor with no development of the telescope eye could be considered as a black metallic Veiltail, while an Oranda which produced its hood rather late in life could make its debut as a Veiltail. Not all such subterfuges are easily exposed, but some are, such as attempts to pass off underdeveloped Veiltails as Fantails. It is perhaps unfortunate that with revision of standards these initial laudable objectives are becoming ever more difficult to achieve.

Another example of the identikit approach is the body standard set for the Bubble-eye, Celestial and Pompon which is in the Nationwide Standards and is essentially the same as that for the Lionhead. This standard is at quite a remove from the bodies commonly seen in these fish. While the Lionhead standard is clearly applicable to Chinese-style Lionheads, it is not an adequate yardstick to judge Ranchu by. The judge must use his discretion, which is what has actually happened and admittedly no great harm has resulted. The bodies of the other dorsal-less varieties in living examples are distinctly different from those portrayed in the standards. Some clarification might be helpful as to whether these standards represented goals for the breeder or guidelines for the judge. A similar situation has arisen in the Fantail; a rather different style of Fantail has emerged in the past 20 years with deeper body and short fins. Perhaps such changes should be monitored and reviewed and changes incorporated when standards are revised. Of course there is the problem that standards may remain unchanged because of a lack of consensus or interest in the variety in question. Postwar changes and developments in the Bristol Shubunkin have been substantial and so changes in standards have been made and readily accepted. The position of each standard and each variety is unique and a pragmatic approach is therefore sensible.

It is one thing to devise sets of standards on the basis of laudable principles but it becomes very difficult with the passage of time to stick to them. Perhaps we should accept the American GFSA viewpoint and regard standards more as guidelines than as sets of rigid prescriptions. This approach is not without its pitfalls but it does permit desirable development to occur, as in the Bristol Shubunkin, whereas an over-prescriptive approach would have stopped it dead in its tracks. Sets of flexible guidelines rather than rigid standards in the hands of experienced and enlightened judges are the best way to ensure progress and avoid stagnation in goldfish culture.

One area regarding standards and guidelines for judging in which consensus has emerged in Britain and America is an agreed Points Schedule. In this the allocation of points is equal between

(1) Body characters
(2) Finnage
(3) Colour

(4) Special characteristics and
(5) Condition and deportment.

So far we have covered the first four aspects and it is perhaps fitting to conclude with a discussion of the questions of condition and deportment.

Condition and deportment

The importance of condition and deportment in show fish is often underestimated. In the current British Nationwide standards 20% of total points are allocated to these aspects. They are of course interrelated, for its condition will obviously affect the way a fish carries itself, i.e. its deportment. As the standards say, fish 'should always be in prime condition, appearing well nourished and free from scars, torn or bloodshot fins, growths or missing scales. Diseased or deformed fish may be disqualified'. It is by no means unknown for fish to receive 'treatment' to enhance apparent condition, an example of which is to strip off much of the mucous coat to enhance the visible quality of colour by use of chemicals. The ethics of such practices are highly questionable in both attempting to deceive judges and also in exposing the fish to attacks of pathogens and parasites by depriving the fish of its first line of defence against them. It is obviously preferable to enhance condition by natural means, for example, enhancing colour of fish by keeping them in the summer in a pond exposed to moderate sunlight. Some advocate green water as a condition enhancer. Water quality is clearly very important indeed as a determinant of condition. Water of the best quality supplied to the domestic user comes from natural aquifers (boreholes) and watercourses where it has not been recycled. Thames water is notorious for being recycled repeatedly; the danger to the fish keeper lies not only in the concentration of nitrogenous compounds but also in the accumulation of metallic ions and the use of additives such as chloramine and pyrethrins to control micro-organisms and crustacea, for example.

While condition and deportment are clearly related in terms of the health and vitality of the individual there is an aspect of deportment which is related to its genetical constitution and the phenotype it has produced. Far Eastern judges of fish in competitions take a holistic rather than the more analytical approach of the Western judge. The fish may be very much more that the sum of the component parts measured on some numerical scale. While this can lead to objective evaluation of fish, the assessment of the overall picture is very much more subjective. The visual impact of an outstanding fish may be immediate and require no totting up of points totals. It is essential that the judge has some leeway to set a seal of approval on particularly fine specimens. The Japanese in particular place considerable emphasis on the way a fish swims. One of the most obvious defects in swimming is that of 'head standing' when the fish has difficulty in maintaining a horizontal position when motionless. In the course of gentle movement there should be no difficulty in maintaining a straight and level course. The movement of the best Veiltails is pure poetry in motion and the epitome of grace and elegance. This does not always result from a syndrome of individually desirable features. The quality of movement should be assessed both laterally and from above. Before the advent of glass vessels evaluation of fish had to be from above when fish were kept in ceramic and latterly enamel bowls. It is important to see fish from both viewpoints, something which is not easy at the present time when exhibition tanks are set up in several tiers.

In a points system of judging it is important that the judge uses discretion in awarding 'condition and deportment' points to recognise the fish as an entity and not just a set of components. Obviously breeders, whether professional or amateur, take due note of show standards generally and condition and deportment in particular. For the amateur interested in competition, the behaviour of a fish on the show bench is particularly important. One of the hazards that a judge faces is that the same fish may conduct itself very differently while being judged and subsequently. Many a judge has been mortified to find that the sprightly, active fish he judged only a matter of hours previously has got the hump and is resting rather disconsolately on the bottom of the tank with fins firmly clamped, looking a picture of misery. It therefore behoves the exhibitor to prepare, train and if possible coach their fish to take in their stride and adapt to changes in water conditions they may encounter at shows. For the commercial breeder deportment is important; lively, active fish are more likely to attract customers than sluggish or lethargic oncs. Both amateur and professional need to ensure that everything possible is done to enhance condition and improve deportment in order to achieve their rather different ultimate objectives.

Goldfish appreciation – a postscript

In the final analysis, whether we like a particular goldfish or not is a matter of personal preference. It is possible to like goldfish in general but to be averse to some of the more extreme varieties. In those varieties which do not arouse negative responses, personal taste may well determine whether we prefer Shubunkins to Fantails and whether we prefer Bristols to Londons. It is possible for personal preference to extend to the minutiae of colour shades and balance and overall patterning. What standards do in effect is to establish common ground and a basis for communication between breeders and to establish common goals. It is also desirable that such communication leads to exchange of stock and to the maintenance and perpetuation of high standards over as wide a range of varieties as possible. With the recent influx of new and exciting variants from China we need to do some soul-searching. We certainly should not abandon what we have which has been developed during the past century but we should not disregard what we may be seeing for the first time. We need perhaps the capacity to persist with what we have been doing and yet be receptive to new ideas stimulated by the novel variants we have seen of late. This could well lead to the revitalising of varieties which are currently at a low ebb and stimulate a new wave of enthusiasm for the goldfish.

Chapter 6
Basic Genetic Principles for Goldfish Breeders

Genetics as a scientific discipline was very slow to emerge in spite of the understandable interest humans have in their own heredity. While the Greek philosophers showed astonishing scientific insights in the physical realm their speculations in heredity were wide of the mark and it is only since the eighteenth century that the effect of their unfortunate legacy has been shaken off. This was due to the work of plant and animal breeders of that time who were instrumental in initiating development of modern livestock breeds and cultivated varieties (cultivars) of useful and ornamental plants.

The middle of the nineteenth century saw a revolution in biological thinking with the publication of Darwin's *Origin of Species*. This was closely followed by the publication of Mendel's classic paper in which he established the basic principles of transmission genetics. His solution of the riddle of heredity, which had previously defeated some of the most brilliant human intellects, was, as is well known, greeted at the time with a totally uncomprehending silence. It is a tribute to Mendel's genius that unlike his contemporaries he was able to circumvent the obstacles to understanding, arising from current deficiencies in basic biological knowledge, which took another 30 years and more to overcome. After the discovery of Mendel's work a century ago, it became possible to develop a comprehensive basis for the understanding of heredity in what is now called the chromosome theory of heredity.

The chromosome theory of heredity

The greatest obstacle to the acceptance of Mendel's ideas can be seen in retrospect to have been the lack of knowledge of cell structure and function. The cellular nature of living organisms was not established until the latter part of the century which was followed by the identification of the nucleus and the discovery of chromosomes. Study of chromosomes showed that their number was constant and consistent for the species in which they had been identified. It was also suggested that the nucleus was the seat of the hereditary function when the processes of cell division were observed. It was found that there were two different types of cell division: the commonest, mitosis, occurred in the process of growth and development which involved replication of cells in the production of the individual and all its tissues and organs from the original fertilised egg cell or zygote. In the course of this division each chromosome was observed to replicate and divide along its length and the daughter chromosomes separated, daughter nuclei were constituted and division of the ambient cytoplasm and cells was completed. The longitudinal pattern of replication had important implications, suggesting that there was an important longitudinal differentiation of the chromosomes and it was necessary that this be perpetuated in the daughter chromosomes and cells. This probably lay at the basis of the suggestion that the chromosomes had an hereditary function.

The second type of division, meiosis, was found to occur only in the reproductive organs of animals and plants. In this there are actually two nuclear divisions but only one division of the chromosomes with the result that four nuclei result, each with only half the original chromosome number. The occurrence of this reduction division means that when eggs are fertilised by the sperms the original chromosome number is restored. Each gamete (egg or sperm) contains the haploid or gametic number of chromosomes which constitutes a complete genome, the genetic blueprint. When the gametes fuse in the zygote, the new individual produced has the genomic set in duplicate. Each chromosome in the egg has a similar partner in the sperm; these are termed homologous. Another important feature of meiosis is that in the first division the homologous chromosomes come together, pair and exchange equivalent segments of their chromosomes. Full accounts of these two types of cell division are presented in the basic genetics texts such as Klug and Cummings (2000) or Weaver and Hedrick (1997).

Mendelian inheritance

Mendel's primary concern was to elucidate the pattern or patterns of transmission of heritable traits from parents to offspring. His task was first of all to identify inherited character differences with the simplest genetic cause. These he discovered in peas where he found contrasting characters in pairs which behaved as identifiable units that could be traced from parents to their progeny over several generations. He was able to show that heredity had a particulate base; his hereditary factors retained their identity after passage through several generations. It was also duplicate in nature: for every factor supplied by one parent there was a corresponding (i.e. homologous) factor supplied by the other. These factors we now call genes and the different forms manifested by particular genes are its alleles, which may be two (the situation considered by Mendel) or more. Mendel confirmed the conclusion of his predecessors that reciprocal crosses gave equivalent results, that is the direction in which the cross is made is immaterial. His first original principle is that when the determinants of two contrasting characters are combined in a hybrid, they can be recovered from the progeny of that hybrid, they segregate and retain their integrity. His second principle, that of independent assortment, is that the factors controlling different characters in an individual segregate independently.

Mendel's work was a beginning and after its rediscovery a century ago a number of apparent exceptions to his laws came to light, most notably that of linkage where parental combinations of characters did not recombine freely. The explanation of linkage had to await the formulation of the chromosome theory. Mendel's genes which underwent independent assortment are located on different chromosomes which recombine freely in meiosis. The discovery of linked genes in sweetpeas by Bateson and his co-workers was explained by Sutton who postulated that such genes (i.e. their loci) were located on the same chromosome.

The development of the Mendelian principles of heredity can best be considered in the light of the models derived from the studies of Mendel and his successors. The selection of the experimental materials was made in all probability after much preliminary investigation of their suitability for the purpose. Peas were eminently suitable for the purpose but *Hieracium*, suggested by Nägeli, clearly was not. Characters selected for study also had to be suitable for the purpose, which those Mendel chose were, that is, they were the simplest differences available which produced clear-cut effects. Mendel's reasoning in making his choices was better focused than that of his predecessors who were interested in studying very complex differences, which defeated their attempts at analysis.

Monohybrid inheritance

Mendel studied seven characters of pea which behaved in a consistent fashion and from which we can derive our basic models. We can then consider the refinement of these and their further elaboration as far as they affect the goldfish. The choice of the pea by Mendel was absolutely inspired; it is self-pollinated and breeds true generally, and numerous characters suitable for study could be identified easily. In exploring the heredity of animals such as the goldfish, breeding tests have to be carried out first to determine whether the character in question is in fact fixed. Logistically some of Mendel's crosses were extraordinarily efficient, especially those using characters of the embryo such as colour (yellow versus green) and surface (round versus wrinkled). He was able to see and score these characters in the seed without actually having to grow them on, which he had to do when characters such as flower colour were studied.

His crossing experiment of varieties with yellow and green cotyledons respectively can be summarised as follows:

Parents	Yellow cotyledons (*P1*) × Green cotyledons (*P2*)
F1 (First filial generation)	All with yellow cotyledons
Self-pollinated	
F2 (Second filial generation)	6022 yellow and 2001 green

The results of all seven different crosses Mendel carried out in this way were that the character of one parent only was expressed exclusively in the first generation (F1). This member of the pair he termed 'dominant'. When F1 plants were allowed to self-pollinate (or self) and produce the second hybrid or F2 generation, both parental characters (or phenotypes) were expressed in progeny. The character which reappeared in F2 was called 'recessive'. Another consistent feature was that approximately three-quarters of the F2 progeny showed the dominant phenotype and one-quarter expressed the recessive character. He made his crosses reciprocally and found that the direction of the cross was immaterial.

The occurrence of dominance had been something of a stumbling block to Mendel's predecessors and contemporaries who were puzzled by characters which appeared to skip generations. This was a problem that Mendel tackled successfully by taking progeny of controlled crosses through several generations and making test crosses. These aspects are best considered in relation to the general model Mendel developed. This is based on a set of premises which he developed on the basis of his own work and that of his predecessors, namely:

(1) Reciprocal crosses gave identical results, that is to say that male and female contributions to the hereditary make-up of their progeny are equal.
(2) The determinants of heredity are particulate and duplicate, that is, they have a stable physical basis and that for every factor contributed by one parent there is a comparable contribution from the other.
(3) These determinants (genes) may exist in alternative states (i.e. alleles) which bring about different expressions of the characters they control.
(4) Fusion of parental reproductive bodies is a random process.
(5) Hybrids between two different parents could transmit to individual progeny either the factor derived from the female parent or its counterpart from the male but not both. (Note: this implies the existence of meiosis in formation of reproductive bodies or gametes.)

Mendel symbolised his factors by letters: upper case (capitals) signified the dominant alleles (e.g. *A*) while lower case (*a*) represented the recessives. He carried out crosses reciprocally and showed similarity of outcome in F1 and F2.

Mendel's monohybrid crossing scheme:

Parents	AA	×	*aa*	*aa*	×	AA
Single factors transmitted to progeny in gametes	*A*	×	*a*	*a*	×	*A*
First generation hybrid F1 producing male and female gametes		Aa			*aA*	

		♀ Gametes	
		A	*a*
♂ Gametes	*A*	*AA*	*Aa*
	a	*Aa*	*aa*

→ $1AA:2Aa:1aa$

With complete dominance *AA* is indistinguishable from *Aa* and so we have progeny with different genetic constitutions (genotypes) which have a common physical appearance (phenotype). This converts the $1AA:2Aa:1aa$ ratio into a $3A\text{-}:1aa$ phenotype ratio where the hyphen can represent either *A* or *a*.

Mendel carried out breeding tests on A-type progeny. When he backcrossed to the dominant parent (*AA*) all progeny showed the dominant phenotype. The cross to the recessive parent either gave a progeny with a uniform dominant phenotype or a progeny with half dominant and half recessive phenotypes. The latter were twice as frequent. He was able by this progeny test with the recessive parent to show that two-thirds of the F2 progeny showed a hybrid or heterozygous genetic constitution and the remaining one-third were not hybrid and potentially true breeding i.e. homozygous.

Mendel's backcrosses:

	AA	×	*aa*	Aa	×	*aa*
Gametes	*A*	×	*a*	*A* and *a*	×	*a*
Progeny		all	*Aa*	1*Aa*	:	1*aa*

This procedure showed that the 3*A*- : 1*aa* segregation was in fact 1*AA* : 2*Aa* : 1*aa* modified by dominance.

Monohybrid inheritance studied by Mendel gave rise to his first principle, which can be stated as follows:

> 'Hereditary characters of organisms are controlled by unit factors which may exist in two or more forms; these can be combined in hybrids and recovered from them in proportions expected on a basis of simple probability.'

Dihybrid inheritance

The study of dihybrid inheritance led to the formulation of Mendel's second principle, which can be stated simply – different pairs of alleles segregate independently, or genes at different loci segregate independently. This was a logical development of the study of single-factor segregation. Perhaps the most simple demonstration of this principle is given by considering the inheritance pattern of two seed characters of pea, yellow vs green and round vs wrinkled. Considered separately Mendel found that both pairs of characters gave 3 : 1 segregations. The two pairs of character differences can be combined so that it is possible to cross a yellow-round stock with a green-wrinkled stock. As expected the F1 showed both dominant characters and was yellow and round. In F2 four classes of progeny were produced those which showed the parental combinations i.e. yellow, round and green wrinkled and those which showed new character combinations yellow, wrinkled and green, round. If the characters were scored separately, 3 : 1 segregations were produced. If they were scored sequentially, first into yellow and green classes and then each class separately into round and wrinkled classes, it was found that within each colour class, yellow and green, there was a 3 : 1 segregation for round and wrinkled; this result can be summarised as below.

Dihybrid segregation

First segregation:

$\frac{3}{4}$ yellow
- $\frac{3}{4}$ round giving $\frac{3}{4} \times \frac{3}{4}$ yellow round — 9/16
- $\frac{1}{4}$ wrinkled giving $\frac{3}{4} \times \frac{1}{4}$ yellow wrinkled — 3/16

$\frac{1}{4}$ green
- $\frac{3}{4}$ round giving $\frac{1}{4} \times \frac{3}{4}$ green round — 3/16
- $\frac{1}{4}$ wrinkled giving $\frac{1}{4} \times \frac{1}{4}$ green wrinkled — 1/16

or an overall 9 : 3 : 3 : 1 ratio.

This result shows that segregation of these two factors is quite independent. A general model of dihybrid inheritance, if we take two pairs of alleles *A/a* and *B/b* and set up a cross similar to that of the monohybrid model, can be set out as follows.

Mendel's dihybrid crossing scheme:

Parents	*AABB* × *aabb*
Gametes	*AB* × *ab*
F1	AaBb
Gametes	

		♀ *AB*	*Ab*	*aB*	*ab*
	AB	*AABB*	*AABb*	*AaBB*	*AaBb*
	Ab	*AABb*	*AAbb*	*AaBb*	*Aabb*
♂					
	aB	*AaBB*	*AaBb*	*aaBB*	*aaBb*
	ab	*AaBb*	*Aabb*	*aaBb*	*aabb*

The result is 1*AABB* : 2*AABb* : 1*Aabb* : 2*AaBB* : 4*AaBb* : 2*Aabb* : 1*aaBB* : 2*aaBb* : 1*aabb* i.e. a 1 : 2 : 1 : 2 : 4 : 2 : 1 : 2 : 1 segregation of genotypes. When dominance occurs as in Mendel's crosses we have four distinguishable phenotypes produced which we can represent as follows:

	Phenotypes	*A-B-*	*A-bb*	*aaB-*	*aabb*
comprising	Genotypes	*AABB, AABb, AaBB, AaBb*	*AAbb, Aabb*	*aaBB, aaBb*	*aabb*
	Proportion	9/16	3/16	3/16	1/16

Modified genetic ratios – gene interaction

Inter-allelic interactions

We have already considered the modification of fundamental genetic ratios by dominance how the genotypic 1 : 2 : 1 ratio becomes 3 : 1 and the dihybrid 1 : 2 : 1 : 2 : 4 : 2 : 1 : 2 : 1 becomes the 9 : 3 : 3 : 1 ratio.

Dominance is the result of the interaction of two different alleles at the same locus. While dominance is of frequent occurrence it is not invariably shown in heterozygotes. These may show a phenotype intermediate between those of the parents (there is a good example in goldfish), which is incomplete dominance. This is not to be confused with co-dominance when different alleles at the same locus produce independent effects, as seen in human blood groups such as the ABO and MN systems where each allele independently produces a distinctive antigen.

Lethal genes have been reported from carp; they are recessive as regards lethality but have a dominant phenotypic effect. These are commonly detected when it proves to be impossible to produce the two different homozygotes expected from matings of heterozygotes. The result of heterozygote matings is a modified 1 : 2 : 1 segregation in which one class is unviable and so it becomes a 2 : 1 segregation. Although mirror scaled goldfish have been produced, presumably from crosses with mirror carp, their segregation patterns have not been studied. The *N/n* locus responsible for the character is a recessive lethal in carp but this has not been tested in goldfish.

Interactions between loci

Interactions between loci are of two kinds, additive and epistatic.

Additive interactions occur when two or more loci affect the same character. The situation is often encountered in connection with pigmentation in organisms as diverse as wheat and humans. Davenport proposed a model for melanin production in human skin, suggesting that two loci were involved with two alleles *P1/p1* and *P2/p2*. His basic premises were:

(1) Although at different loci, effects are similar and approximately equal.
(2) Alleles are either active *P1/P2* in producing pigment or inactive *p1/p2*.
(3) They do not show dominance.
(4) Their effects are additive.

The situation can be modelled theoretically. A mating between *p1p1p2p2* and *P1P1P2P2* individuals would produce an intermediate F1, *P1p1P2p2*. If two such individuals were to mate, expected frequencies of individuals carrying 0, 1, 2, 3, 4, active alleles would be 1 : 4 : 6 : 4 : 1, which gives a symmetrical distribution a simple bell-shaped frequency distribution. This is the segregation pattern which underlies the inheritance of quantitative or metrical characters in contrast to the qualitative characters such as those colour differences which show Mendelian segregation patterns. It must be emphasised that in both gene types, transmission from parent to offspring is according to the Mendelian model.

Epistasis occurs when two separate loci are concerned with the production of phenotypic differences in an interactive fashion. One locus controls whether the other produces its effect or not; the controlling locus is termed epistatic, the locus controlled is hypostatic. The examples studied so far showed dominance and so what we are concerned with is the way that the basic 9*A-B-* : 3*A-bb* : 3*aaB-* : 1*aabb* ratio is modified. This new modification arises when two or more classes share the same phenotype. A range of examples is considered below.

(1) If both *A* and *B* are required to produce a specific phenotype a two-class 9 : 7 segregation will be produced; none of the 3 : 3 : 1 classes have both the essential genes. In this particular case epistasis is reciprocal and recessive.
(2) If the action of *A* and *B* is sequentially related (*A* produces an identifiable intermediate state but *B* is required to produce the final state), the segregation produced will be 9 : 3 : 4. The largest class will have both *A* and *B*, the smallest will have only *A*, which is absent in the remainder. *B* on its own produces no effect. In this case the *A* locus is epistatic to *B*; the homozygous recessive exerts the effect and this is recessive epistasis.
(3) A third example is seen when one locus, *A*, affects whether the action of *B* can be detected. If *A* produces a black seed pigment, for example, this may obscure pigments produced by the *B* locus; if *B* and *b* were to produce different pigments a 12 : 3 : 1 segregation would be produced. The locus *A* shows dominant epistasis.

At least one epistatic interaction has been reported in the goldfish.

Linkage

Linkage was discovered in a dihybrid cross progeny which Bateson and his collaborators produced in 1905. In a cross between two sweetpea varieties, one having purple flowers and long pollen grains and the other red flowers and round pollen grains, they obtained the following results:

Parents Purple long × Red round
F1 all Purple long

F2	Purple long	Purple round	Red long	Red round	
Expected ratio	9	3	3	1	
Expected frequencies	240	80	80	27	Total 427
Observed frequencies	296	19	27	85	Total 427

Clearly the dihybrid segregation does not agree with expectation; however, if one examines the two monohybrid segregations they do.

Expected ratio	3 : 1
Expected frequencies	320 : 107, total 427
Observed frequencies:	
purple versus red	315 : 112
long versus round	323 : 104

It was on the basis of observations such as these that it was accepted that Mendel's principle of independent assortment was not universally applicable. Sutton suggested that this association of parental characters (or linkage) occurred because the loci concerned were on the same chromosome. Recombination of these was possible as a result of chiasma formation (crossovers) in meiosis. The frequency of recombination was proportional to the extent of physical separation of the loci in the chromosome. Loci which were close would combine less frequently than those which were distant. Loci at opposite ends of a chromosome might well recombine as frequently as if they were unlinked because a chiasma would be expected always to form between them.

By measuring recombination frequencies between genes it is possible to produce genetic maps of the chromosomes. This has been done most notably in the Drosophila fruit-fly which has four pairs of chromosomes and in which linkage of genes is frequent: on average there is a 25% chance that any two genes will be linked. In the case of goldfish with 50 pairs of chromosomes there is only a 2% chance. As a result linkage studies and gene mapping have not been attempted. But linkage studies are not without significance in the goldfish. The greatest practical significance of linkage in goldfish breeding and selection is related to linkage drag. When conscious selection is carried out, not only are the genes selected established in progeny populations but also others which are closely linked to them. These may be disadvantageous or deleterious and if the selected trait is fixed genetically so also may closely linked genes with possible detrimental consequences. It may be possible to overcome this problem by providing opportunities for genetic recombination to occur by judicious crosses and careful subsequent selection. Recurrent selection of this type may keep the problem at bay.

Sex linkage

Sex linkage provides an exception to the basic Mendelian premise that reciprocal crosses are equivalent. Sex linkage occurs in those organisms in which a chromosomal mechanism of sex determination occurs. In this a pair of chromosomes are concerned with the operation of a developmental switch which controls the direction of sexual development towards maleness or femaleness. It is thought that initially this switch operated through allelic differences but that progressively this evolved into linked co-adapted gene complexes located on one or other of a pair of homologous chromosomes. These homologous sex chromosomes diverged not only in their gene content but also in their morphology, becoming recognisably distinct *X* and *Y* chromosomes. The operation of the system is such that *XX* and *XY* chromosome constitutions produces different sexes, commonly *XX* being female and *XY* the male configurations. There are cases, in birds and butterflies, where females are *XY* and males *XX*. In reproduction the *XX* individual produces only *X*-bearing gametes (homogametic) while in the *XY* individual these may carry either the *X* or the *Y* chromosome (heterogametic). In mating, equal numbers of *XX* and *XY* zygotes are produced and a 50 : 50 sex ratio of males and females is established.

Genetic divergence between *X* and *Y* chromosomes occurs and the gene content can become very different. The *X* and *Y* have a common homologous segment which enables them to form a meiotic bivalent. The non-homologous or differential segments of the *X* and *Y* chromosomes commonly contain genes which have no counterpart in their opposite number, leading to *X*-linked and *Y*-linked genes, which are well exemplified in many of the male guppy's colour genes. The typical *XX* female is basically diploid or disomic in its sex chromosomes constitution; the *XY* male is haploid or monosomic as regards the *X*. Transmission of the *X* chromosome in the female is equally to the male and female progeny but in the male the *X* passes to female and the *Y* to male progeny. This produces a markedly different transmission pattern to that of genes located in the other chromosomes (the autosomes). *X*-linked genes in *XY* males will all be expressed since dominance will be inoperative while in females normal dominance can occur. *Y*-linked genes are transmitted exclusively in the male line.

Sex linkage is an important feature of the genetics of many tropical ornamental species, most notably the guppy, (Schröder 1976) but so far no significant features of the goldfish express it. Yamamoto (1975) has produced experimentally a *YY* male and earlier (Yamamoto & Kajishima 1968) had studied the heterogamety of male goldfish.

Polyploidy

Polyploidy arises when individuals are produced which carry more than the normal genome complement, that is, more than the customary pair of genomes. In the carp family, the Cyprinidae, to which the goldfish belongs, the basic chromosome complement is 50; this is the somatic complement. It is made up of two sets of chromosomes which are homologous (that is, they have a common evolutionary origin), one which is of maternal and the other of paternal origin. Such a gametic chromosome complement constitutes the genome. In the goldfish the chromosome complement is 100, which is double that of the majority of Cyprinid fishes (and of the bony fishes in general). The means by which such doublings of the chromosome complement can come about are well understood and they arise in the course of mitotic and meiotic cell divisions by failure of the chromosomes to separate. This can produce somatic cells with double the normal chromosome complement, or reproductive cells in which the somatic chromosome number has been maintained. The fusion of two such gametes will produce an individual with double the normal chromosome number.

Autopolyploidy and allopolyploidy

There are two distinct types of polyploid: firstly, those in which all the genomes present are closely similar (the autopolyploids) which we can represent by the genomic formula *BBBB*; the second type (the allopolyploid) combines two distinctly different genomes which we can call *B* and *C*, which therefore has the genomic formula *BBCC*. These two types have different origins. Autopolyploidy originates when in a dividing cell, spindle formation fails to occur, daughter chromosomes fail to segregate and the nucleus is reconstituted with double the normal chromosome complement. Such a cell could produce a polyploid sector which if it were to give rise to reproductive tissue could generate polyploid progeny. Allopolyploidy arises as a consequence of hybridisation between species which are sufficiently closely related to produce viable hybrids but whose genomes have diverged as a result of chromosome structural change and differentiation at the gene level. As a result of such changes the F1 interspecific hybrid may be completely sterile or of reduced fertility. Meiosis in such hybrids is irregular, and may result in abortion of the process and failure of the reduction division (meiosis I) so that, if the second division is completed, two gametic cells result with an unreduced chromosome complement rather than the normal four with a reduced number. The fusion of such unreduced gametes can produce progeny which combine pairs of genomes from the two parental species, making normal meiosis possible with the restoration of normal fertility.

The success of meiosis depends on the regular pairing of homologous chromosomes. This is controlled by three factors: the first is the congruence of the gene (loci) sequences of the potential pairs, the second is the frequency of chiasmata in the chromosome pairs or bivalents (at least one chiasma per bivalent must form), and the third is chromosome length, which is important in autopolyploids. When chromosomes are short and the chiasma frequency is low, (about one) bivalent pairing will predominate and a normal meiosis will ensue. Long chromosomes and higher chiasma frequencies (two or more) will permit multivalent formation and irregular segregation in meiosis. Such segregations can give rise to aneuploidy, that is loss or gain of chromosomes, reduced fertility and poor viability of progeny. In such instances autopolyploids suffer reduced fertility.

Genetic consequences of polyploidy

The genetic consequences of polyploidy depend on the cytogenetic nature of the polyploid, that is, whether it is an allopolyploid or autopolyploid. Suppose we have a locus *A* with alleles *A* and *a* in both situations. In the allopolyploid, if we have two genomes *B* and *C* we might well find that the *A* locus is present in both genomes. However, due to the divergence between the two genomes it could well be that the homologues carrying *A* in the *B* genome would pair preferentially with those of the *B* genome and similarly those in the *C* genome. The upshot of this is that we would have a pattern of duplicate inheritance if we differentiate the *A* loci in the *B* and *C* genomes as *A1* and *A*, our expectation from the cross *A1a1Aa* × *A1a1Aa* would be 9*A1-A-* : 3*A1-aa* : 3*a1a1A-* : 1*a1a1a*a (where the hyphen may signify either *A1* or *a1*, *or A* or *a*).

With equivalence in the dominance of *A1* and *A* we would obtain a phenotypic frequency of 15 *A* (*A* and *A1* combined) to 1 *aa* (*a* and *a1* similarly combined).

In the autopolyploid the situation is rather different because any of the four chromosomes carrying the A locus could be combined with any of the remaining three homologues. We can assume normal bivalent pairing (if chiasma frequency is about one) and normal pairing in Meiosis I and normal chromosome separation in both divisions. We then have the possibility in the cross *AAaa* × *AAaa* (all *A* and *a* alleles are in fully homologous chromosomes) of producing the following array of gametic types in both males and females. If we number the chromosomes 1, 2, 3 and 4 we can produce the following gametic array.

Chromosome	1	2	3	4
Allele carried	*A*	*A*	*a*	*a*

Possible gametic combinations

	4(a)	3(a)	2(A)
1(A)	Aa	Aa	AA
2(A)	Aa	Aa	
3(a)	aa		

Frequencies 1*AA* : 4*Aa* : 1*aa*.

In a mating of *AAaa* × *AAaa* heterozygotes, our Punnett Square would be:

	1AA	4Aa	1aa
1AA	1AAAA	4AAAa	1AAaa
4Aa	4AAAa	16AAaa	4Aaaa
1aa	1AAaa	4Aaaa	1aaaa

Genotype frequencies 1*AAAA* : 8*AAAa* : 18*Aaaa* : 8*Aaaa* : 1*aaaa* with dominance of *A* phenotype frequencies are 35*A*--- : 1*aaaa*.

In the autopolyploid the frequency of the homozygous recessive is 1 in 36 whereas in the allopolyploid the expectation is 1 in 16, slightly more than double.

Interestingly enough matings between heterozygotes *Aaaa* × *Aaaa* give the same expectation as *A1a1aa* × *A1a1aa* or *a1a1Aa* × *a1a1Aa*, a 3 : 1 segregation of *A-* : *aa* phenotypes.

If the heterozygote *AAAa* × *AAAa* mating is performed, no segregation would be expected as the maximum number of recessive *a* alleles in progeny would be 2. Similar outcomes would be expected in $A1A1Aa \times A1A1Aa$ and $A1a1AA \times A1a1AA$ matings (for further details see Smartt & Bundell 1996). One can conclude that if heterozygotes known to contain equal numbers of the appropriate dominant and recessive alleles gave on mating a 15 : 1 segregation, the inference would be that the constitution of the homologous pairs of chromosomes was allopolyploid; however, if it were 35 : 1 (or close to it) then the inference would be that the chromosomes in question were fully homologous and an autopolyploid condition existed. In such a situation the expected mutant phenotype frequency would be slightly less than half that in an allopolyploid population.

Developmental aspects

Embryology of the goldfish has been studied by Affleck (1960) and Kajishima (1960b) and with the exception of the twin-tail mutation most of the mutants of interest to the fancy goldfish breeder do not make their presence felt at this stage. During the larval phase some differences in the development of chromatophores become apparent and are clearly seen in the fry, notably in the case of the transparent scale mutation. The majority of mutants develop their phenotypic characteristics rather later and progressively; good examples are depigmentation and development of the hood in the Lionhead and Oranda where the characters may develop early in a few months or take years. In the case of the twin-tail the die seems to have been cast during embryonic life although its full development may not be apparent until larval development has been completed and the fry phase has been initiated (Fig. 1.3a, 1.3b, 1.3c).

In the case of the twin-tail and the dorsal-less conditions, the future development of finnage is indicated by presence, absence or duplication of the embryonic fin folds. Where the dorsal or anal fin is absent this is preceded by the failure of the appropriate segment of the fin fold to differentiate; where anals and caudals are duplicated so is the fin fold in the appropriate areas. In the course of development not only do the fins themselves develop during the larval phase but also the appropriate supporting skeletal structures.

There are a number of development issues which merit discussion, and these relate firstly to the way the genotype responds to the environment and secondly to the way the mutant gene interacts with the other genes in the genotype. The latter for convenience can be called the background genotype. It seems that these interactions can be complex, judging by the range of effects which can be produced on the phenotype.

A common and perfectly reasonable assumption is that the genotype determines the phenotype and in the special case of incomplete dominance each genotype has a characteristic phenotype. In the study of human genetics some apparent exceptions occur to the general rule that an individual heterozygous for a dominant gene expresses the dominant phenotype. One of the most interesting of these is the mutation which produces polydactyly. This feature has been subjected to extensive pedigree analysis and in the great majority of cases behaves as a conventional dominant. Occasionally individuals who do not express the character produce progeny which show it although neither parent does so. This has been a recurring feature of these pedigrees and it has been concluded that the mutant, though present, has for whatever reason failed to express itself. It may be that some trigger controlling the mutant development pathway has not operated and normal development has occurred. In some pedigrees the firstborn although showing normal development has produced polydactylic children, so it is possible that a prenatal stress factor might be implicated as the first pregnancy generally would be more stress-free than subsequent ones.

When polydactylic individuals are examined, it is commonly found that the developmental range of the mutant feature show considerable variation. The number of additional digits produced may be one or two, development of the extra digits may be complete, and fully functional additional fingers and toes may be produced. Development of the additional digits may be incomplete and as little as a splint of bone may be produced. Production of extra digits is not consistent on all limbs. Some may be normal in appearance while the others may show a developmental range. This phenomenon has two aspects: *penetrance*, which relates to whether the mutant phenotype is shown at all, and *expressivity*, which is related to the extent to which the mutant character is developed. Polydactyly is thus classed as an incompletely penetrant dominant mutation with variable expressivity. Two comparable mutations have been identified in the goldfish, the twin-tail and the out-turned operculum, which will be discussed later.

It is important not to lose sight of the influence that the environment may have in the course of development. The effect may appear to be a direct one on the action of the gene itself or the gene product. If a mutation confers temperature sensitivity on an enzyme, say, variation in temperature could influence the phenotypic outcome. If this locus interacted with others there is the possibility of further interactions. This provides a simple illustration of the effect that other genes present in the genome may have. This is related to the time at which the loci in question actually operate. Those which exert their effect early will have a more profound effect on average that those which are activated later. In considering mutations we have to consider not only the physical environment but also their genetic environment.

Quantitative genetics

Quantitative genetics is concerned with the inheritance of characters which are on the one hand metrical in nature, related to weight and dimensions for the most part, and which are under the control of several genetic loci (that is, they are polygenic in nature) as contrasted with qualitative characters under the control of few or even single loci, which are termed oligogenic. The most obvious difference between polygenic and oligogenic segregations is that the latter produce distinct classes of variants while in the former variation is continuous. This difference has historical significance as it was the basis for the controversy between the Galtonians and the Batesonians. The difference was reconciled by Fisher who showed that the more genes that were implicated in development of a specific character the less chance there was of obtaining discrete progeny classes.

The characteristic of a polygenic segregation in F2 is that when a frequency distribution is plotted it will produce a bell-shaped curve, characteristic of the 'normal distribution' in which phenotype expression frequencies will be greatest around the mean and fall away progressively to both extremes. In producing models of polygenic inheritance some simplifying assumptions are made: that the effects of the loci are equal and that epistasis and dominance are absent. Alleles at loci can be visualised as having a positive, negative or zero effect on a basic level of phenotypic expression; the balance of positive and negative factors will determine the actual level of expression of the character in the phenotype. Typical quantitative characters in goldfish are length and breadth of fins, and length and depth of body, which in appropriate F2s do not segregate into distinct classes but are continuously variable.

The analysis of quantitative variation is complicated and can only be outlined briefly. More extended treatments will be found in texts on animal and plant breeding, to which the reader is referred for further detail. What is entailed can be summarised briefly as follows. When a cross is made between two different goldfish varieties and an F2 is produced from the F1, a very complex segregation pattern is produced in which a wide range of variation is expressed for many characters, both oligogenic and polygenic. This variation is quantified statistically as the phenotypic variance, which can be partitioned into two major components, the genetic and the environmental. In turn the genetic variance can be subdivided into three subordinate components: the additive component quantifies differences between homozygotes at loci, the dominance component arises from interactions between alleles in heterozygotes (i.e. it is a measure of the cumulative effects of dominance), and the third, the epistatic component, arises from the effect of interactions between loci. It is useful to note that through sophisticated statistical manipulations it is possible to estimate the magnitude of these components. While it is unlikely that any goldfish breeder would actually wish to undertake these operations it is quite important to know that they are feasible because of their importance regarding the important subject of heritability.

Heritability

To the geneticist heritability is the proportion of the phenotypic variation observed in a segregating population which is due to the genetic component.

$$\text{Phenotypic variance} = \text{genetic variance} + \text{environmental variance}$$

$$\text{Heritability} = \frac{\text{genetic variance}}{\text{genetic variance} + \text{environmental variance}}$$

This little equation sums up very neatly what can be a very complex situation indeed. During a large part of the twentieth century an utterly futile argument raged over which was the more important, nature or nurture, or to put it another way, genotype or environment. The short answer is that both are important; the genotype can be regarded as determining the development potential of an individual while the environment controls the amount of this potential which is realised. A human example might usefully illustrate the point. Individuals commonly vary in height; in an optimal environment the genetic component determines the outcome, under a feeding regime inadequate for normal growth such differences may not exist, in which case the nutritional environment is pre-eminent. Heritability is not necessarily fixed although there are some instances where heritability may be 100%. No amount of feeding or environmental manipulation will alter the eye colour of a white rabbit (Allard 1960, Simmonds & Smartt 1999).

As far as the goldfish is concerned we need to be aware of the concept of heritability, which is all too frequently dismissed with inadequate thought or completely ignored. The breeder ignores the implications of heritability at his peril. What the breeder must do in order to succeed is to select for the highest possible quality of broodstock and then to raise its progeny in the closest to optimal environmental conditions that he can produce. It is a matter of common observation that with the same pair of fish spawning in a succession of seasons under ambient climate conditions, consistent results are not necessarily obtained. Usually a mixture of good, bad and indifferent seasons will be experienced. It is sensible for the breeder to identify critical environmental factors and be prepared to take appropriate action to maintain optimal conditions as far as possible.

Population genetics

Some familiarity with the very basic principles of population genetics is necessary if we hope to understand more completely the evolution of the domestic goldfish. It arose from a wild population from which the domesticated population was segregated and which subsequently has become divided further into sub-populations which we call varieties. The chief feature of these progressive changes is that the effective size of populations has become smaller. The implications of this need to be considered.

The fundamental concept of population genetics is that of the Hardy–Weinberg equilibrium. The model assumes that the population concerned is large and that mating within the population is at random. It is, however, a very robust model and corrections can be made when the initial assumptions for equilibrium are not met, i.e. when populations are small and when mating is not random. In addition other corrections can also be made to fine-tune the model to apply to a wide range of conditions which can apply to populations.

Basic model

This is developed from the consideration of a single genetic locus at which two different alleles, *A* and *a*, occur. Every individual in the population will carry an *A* or *a* allele in each of the homologous chromosomes, that derived from the mother and that derived from the father. There will be in the population as a whole a frequency of the *A* allele which we can call p and a frequency of the *a* allele we can term q. If we set the combined frequencies of *A* and *a* to equal 1 then $p + q = 1$. (Convention dictates that we use 1 rather than 100%, a possible alternative). If we consider then what happens in mating, assuming that the allelic (commonly called 'gene') frequencies are the same in both sexes, then the proportions of eggs and sperm produced carrying the different alleles will both be p and q. The result of random mating between males and females will be $(p + q)$ eggs fertilised by $(p + q)$ sperms. The result of this will be that we have $p^2(AA) + 2pq\,(Aa) + q^2(aa) = 1$ (i.e. comprises the whole progeny) since $p + q = 1$, $(p + q)^2 = 1$. This means that if we know the relative frequencies of the two genes, *A* and *a*, we can predict the frequencies of the progeny genotypes *AA, Aa* and *aa*. An important consequence of this situation is that if we consider the recessive *a*, and its frequency q in the population is 0.1 (i.e. 10%), the homozygous recessive will occur at a frequency of 0.01 (1%) in the population. The greater proportion of the recessive genes will be present in the heterozygotes, and as the frequency q becomes smaller this tendency increases geometrically. With a frequency $q = 0.001$ the frequency of homozygotes will be one in a million and 99.8% of the alleles will be present in heterozygotes and only 0.2% in homozygotes.

We can appreciate from this consideration that with large populations we could accumulate at low frequencies a wide range of mutant genes which might only rarely be expressed. However, if we take small and unrepresentative samples from large populations and allow them to interbreed, the frequency of some rare recessives in the new population may be much greater than in the original, while others may not be represented at all. This is due to sampling error and gives rise to the 'founder effect' which has been studied in human populations where large populations have been established in new territory by small bands of immigrants. Rare and often pathological conditions may become commonplace in such communities. Another characteristic of small populations is the occurrence of 'genetic drift', which occurs when perhaps due to a small and unrepresentative (genetically) group of parents plus sampling error, gene frequencies may fluctuate widely, which can result in fixation of some alleles, even disadvantageous ones. These phenomena have almost certainly been involved in the evolution of goldfish variants and their effects have undoubtedly been reinforced by inbreeding.

The effect of inbreeding is to reduce the frequency of heterozygotes by a coefficient F (Sewall Wright's coefficient). This enables a simple correction to be applied to the Hardy–Weinberg equilibrium equation. Since the reduction of heterozygotes $= F \times 2pq$, the frequency of both homozygotes will each increase by pq. So the corrected equation becomes $(p^2 + Fpq)AA + (2pq - 2Fpq)Aa + (q^2 + Fpq)qq = 1$. A similar correction can be applied for the effects of genetic drift, which need not concern us here. Inbreeding may be incidental when we establish a new population from few parents or deliberate when we mate closely related individuals and even siblings on occasion.

In addition to inbreeding and genetic drift, correction of the Hardy–Weinberg equation can be made to accommodate the following effects: (1) mutation and (2) selection. The Hardy–Weinberg principle can also be applied to more complex systems than the simple single-locus two-allele situation on which the basic equation is based. These are (1) multiple allele systems and (2) sex and sex linkage – where gene frequencies are not the same in the two sexes and where *X*- and or *Y*-linked gene frequencies are considered. For the reader interested in a more detailed account of population genetics the very lucid treatment of Mettler and Gregg (1969) can be highly recommended. We shall return to the question of inbreeding when we consider breeding strategies and methodology.

Breeding strategy

Successful breeding strategies in essence depend on the successful manipulation of inbreeding. This is very definitely a two-edged sword: on the one hand it serves to fix the obvious desirable characters for which selection operates while equally it serves to fix less desirable (and even undesirable) characters for which no selection is being practised. Linkage drag can be a very potent factor in the fixation of undesirable alleles. For this reason it would help if selection pressures could be relaxed periodically, permitting recombination to occur between desirable genes and their nearest linked neighbours. This is the basis of the recurrent selection strategy used by animal and plant breeders in the production of the inbred lines necessary to produce commercial F1 hybrids ranging from chickens to maize.

It was observed early in the twentieth century by American plant breeders that corn (maize) could be inbred for many generations until it became in effect a true breeding line.

Maize is normally out-pollinated, and open-pollinated populations are highly heterozygous. In the production of inbred lines it was noted that there was a progressive decline in vigour and that in fact some lines failed to survive. However, it was found that, when the surviving lines were crossed in all possible combinations, some combinations had their vigour restored. Of these some particular combinations had levels of vigour which exceeded those of their parents. This clearly demonstrated not only that inbreeding depression was reversible but also that if this bottleneck could be negotiated there is a real prospect of a substantial bonus in what was termed 'heterosis'. The latter is a contraction of 'heterozygosis'. This name was coined because inbred lines were thought to be highly homozygous and that heterosis came about when lines which were homozygous for very different sets of alleles were brought together. Naturally the F1 progeny of two such lines would be highly heterozygous. However, heterozygosity *per se* as an explanation of heterosis did not remain unchallenged for long.

The whole field of inbreeding depression and heterosis (or hybrid vigour) is best understood in terms of the genetic structure of populations, which is lucidly covered by Mather (1973), and the application of such knowledge is well covered in a basic way in Allard (1960). Although a plant, maize (since male and female organs are separate) provides useful guidance in the study of animal breeding systems. Study of maize has provided very valuable insights into the genetic nature of hybrid vigour and its causes. The two camps in the controversy became known as the 'dominance' and the 'overdominance' schools. The argument of the first was that heterozygosity in the F1 actually covered up the effects of deleterious recessive alleles and that if these could be removed and the favourable dominants made homozygous it would be possible (theoretically at least!) to obtain a highly heterotic individual which was also homozygous. The 'overdominance' view is that heterozygosity is essential for heterosis. The argument can be illustrated by a model in which we have two alleles, A and $A1$, which perform similar functions but do so in rather different ways, for example, by producing different forms of an enzyme with different temperature optima. The argument has at times been furious and as is so often the case neither opposing view necessarily encapsulates the whole truth. In fact there is room for more hypotheses such as the 'balance hypothesis' in which the heterotic individual is conceived as having better than average balance in the genotype relative to the environment in which it occurs. These hypotheses are not mutually exclusive and all broaden our understanding of what may be involved genetically.

Let us consider the practical application of this knowledge and understanding. A breeder may produce or acquire a single outstanding individual which he wishes to propagate to exploit its genetic endowment. The sensible strategy is to cross the outstanding individual with the best available similar individual to produce an F1. Selected individuals from this F1 can be backcrossed to the high-quality parent to produce the first backcross F1 (BC1F1). The process can then be repeated and a second backcross generation (BC2F1) produced and this process can be repeated further. The genetic implications of this procedure are set out as follows.

Parents	Quality parent P1 (recurrent parent) × Non recurrent P2 parent		Overall mean genotype
	↓		
	P1 × F1	→	50% Recurrent
	↓		
	P1 × BC1F1	→	75% Recurrent
	↓		
	P1 × BC2F1	→	87.5% Recurrent
	↓		
	BC3F1	→	93.75% Recurrent

We can see that in such a backcross programme we can produce after four crosses a progeny population with an average 93.75% of the genotype of the original quality parent. Selection in each generation is possible and we can certainly eliminate individuals which show undesirable characters. We must bear in mind constantly the fact that while the original quality parent was probably highly heterozygous, a great many loci in the backcrosses will become homozygous, the consequences of which can be both good and bad. Positive selection for the good features should obviously be maintained, but careful note should be taken of the deleterious characters which occur generation by generation. Recurrence of the same undesirable phenotype in successive backcross generations would be a particular concern; steps would need to be taken to eliminate this by careful progeny testing of selected lines.

In any strategy involving inbreeding it is necessary to take effective steps to insure against excessive fixation of deleterious alleles. The consequences of this are not always apparent. Obvious signs and consequences of inbreeding depression are the production of high frequencies of deformed and undersized individuals; other consequences may be almost impossible to detect and very insidious in the way they come about. The maintenance of effective selection against genes which depress viability may still permit fixation of alleles which disturb reproductive processes and drastically affect fertility. Some extremely successful strains to my knowledge have petered out because individuals although apparently healthy never spawned and in some instances showed no inclination to do so. The remedy for this predicament is to maintain at least two and preferably more distinct lines with occasional intercrosses to reduce the probability of fixing deleterious genes which reduce fertility. Breeders who have examined *post-mortem* fish which despite their best efforts to spawn them have never done so, have sometimes found that these individuals have failed to develop any reproductive organs.

A practically effective way of maintaining genetic variability within a strain is when selecting parents for spawning to seek individuals whose characteristics are complementary rather than identical or which in metrical characteristics such as length of finnage and body depth span the acceptable range. The objective is to produce variable progeny among which selection can be practised and, it is hoped, desirable phenotypes selected.

These strategies perhaps go against those developed in tropical fish species as set out by Schröder (1976) in which the objective is to develop 'pure lines', that is true breeding homozygous lines. In diploid species such as the guppy which have been exposed to sustained selection over more than half a century, a great many deleterious alleles have been eliminated. The goldfish breeder would like to establish 'pure lines' also but in practice this is much more difficult to achieve in the more exotic varieties, though the late J. Linale did maintained a line of Broadtail Moors which bred true. This is exceptional and the difficulty posed by the tetraploid genetic system of the goldfish and the possibility of tetrasomic inheritance of some deleterious genes at least, makes their elimination very difficult if not impossible to practice.

Breeding plans such as those outlined by Schröder can be studied usefully by goldfish breeders in order to formulate breeding strategies, provided that the hazards outlined are taken into account along with the other peculiar features of the genetic system of the goldfish.

Chapter 7
The Genetics of Goldfish Variety Evolution

The historical sequence of events in the development of the range of goldfish varieties is reasonably clear; we are now in a position to attempt a genetic analysis of what has actually transpired. By way of summary of what we need to analyse, we can recapitulate the crucial events which occurred in the process. Undoubtedly the initial and most critical event was the occurrence and establishment of the xanthic form of the goldfish in domesticated populations. This was accomplished in the Song period. The second genetic change which occurred was the establishment of the twin-tailed variant in the latter part of the Ming period. Although the establishment of the xanthic mutation was the *sine qua non* of domesticated goldfish variety development, it was this twin-tail which gave it its unique stamp. Many species of fish have been domesticated as ornamentals in the past century but in this major characteristic (and some minor ones) it remains unique.

A logical treatment of the whole evolutionary genetic process should give primary consideration to the quintessential features. Discussion of numerous secondary but nonetheless important characters which have developed in profusion and which have contributed greatly to the appeal and attraction of the goldfish, will follow.

The xanthic mutation

Xanthic mutations are almost a commonplace among ornamental fish, running the gamut from Angels to Zebras. However, the xanthic mutation in the goldfish shows a number of unusual features. In the first place the character is dominant and in the second place it comes about in juvenile fish from the age of about two months but may be delayed even to adulthood and may not occur for several years.

The genetic control of this feature was not easily resolved and it was not until 1977 when Kajishima analysed the colour question that a satisfactory explanation emerged. Kajishima identified a pair of genetic loci at each of which there were two alleles *Dp1/dp1* and *Dp2/dp2*. The dominant alleles *Dp1* and *Dp2* produced loss of melanophores from the age of approximately two months onward and demelanisation could be produced by the two loci acting individually or together. Kajishima ascribed the wide range of the time scale over which this depigmentation could occur to a dosage effect; earlier depigmentation could be expected to occur in individuals which carried four active alleles than those with three, two or a single such allele. Late depigmentation could be explained on the basis of possession of only a single active allele. By virtue of this, positive selection for early depigmentation would be (naturally) in favour of the homozygote, as might be expected in the fish-farming situation. In a feral goldfish population, although selection would be against the conspicuous, brightly coloured individual, the depigmentation gene could well persist in late-decolouring individuals which might well have reproduced before they lost their melanophores. In turbid waters selection against highly coloured individuals in feral populations could be reduced, which might explain the surprisingly high numbers of brightly coloured feral goldfish found in the Lake Erie basin by Mr Al Thomma (personal communication).

Kajishima has therefore resolved some very perplexing anomalies in the way the xanthic mutation behaves in the goldfish. He has also been the first to come to terms with the fact that the goldfish is polyploid in the genetic analyses he has carried out. In this he has followed and developed the work of Ohno (1970) in the specific context of the goldfish. On this hypothesis then the loci *Dp1* and *Dp2* are homologous and indistinguishable in their effects. Because the individual goldfish chromosomes are small, the probability is that the duplicated chromosome complement produces bivalent pairing in meiosis; however, it does raise the possibility of multivalent pairing which could, on perhaps rare occasions, produce unexpected results. An example could be the sudden appearance of wild-type progeny in an apparently true-breeding line. It is by no means certain whether the goldfish is autopolyploid, an allopolyploid or a segmented allopolyploid (i.e. an intermediate between the first two types). If it is an autopolyploid then the *Dp* alleles would be either *Dp* or *dp*. Only if it were allopolyploid (or perhaps a segmental allopolyploid) is the situation best represented by Kajishima's *Dp1/Dp2* notation. Although the indications are that cytogenetically, that is in terms of meiotic behaviour, the goldfish genome is diploidised, it is not clear to what extent this applies at the level of the locus. In general it can be said that 15 : 1 segregations indicate that diploidisation has not occurred whereas 3 : 1 (and of course 1 : 2 : 1) segregations indicate that it has.

Variation in the timing of depigmentation has been noted. In the majority of cases, in early depigmenting stocks, the colour change occurs in the first summer. The minority which do not change in their first year may very well do so with the onset of warm weather in the second. A similar situation can arise with late spawnings; cooler conditions in autumn may fail to trigger the depigmentation response.

The mutant is advisedly called the depigmenting gene because although the cell death of the melanophores is the primary consequence of its action, on occasions the other type of pigment cell, the xanthophore, may be destroyed to a greater or lesser extent, producing red and silver (or white) variegation or completely silver fish. Variegated fish, often called Sarassas (or Tanchos when there is a large red spot on the head), are very acceptable but silver fish which have completely lost their pigments are of much less value. It is a matter of common observation that those fish which depigment very early are the most likely to lose their pigment completely. It is thought, in these cases, that the melanophage cells which consume the melanophores may, when these are disposed of, turn their attention to the xanthophores. It is possible that with the passage of a little more time, the xanthophores are resistant to this action. There may be some constituents of the genotype (the background genotype) which control and modify the action of these phage cells and bring the process to a halt when only the melanophores have been removed. It could also be a gene dosage effect.

As has already been noted the xanthic mutation in the goldfish is unusual when compared with similar mutations in behaving as a dominant. We must bear in mind the fact that the xanthic character has been selected for positively for 1000 years and perhaps longer. Fisher (1930) demonstrated that in the domestic fowl dominant mutations behaved as recessives in the genetic background of the jungle fowl progenitor. This was shown by crossing domesticated and wild-type fowls, followed by repeated backcrossing to the jungle fowl while maintaining at the same time selection for the mutant character. In order to ascertain whether a similar change had come about in the goldfish, Smartt (1999) crossed a female red metallic Veiltail with male Crucian carp (*Carassius carassius*); all the progeny retained the wild-type coloration for five years of close observation and none depigmented. It would appear that the contribution of the Crucian carp parent to the background genotype was sufficient to cause the mutant alleles to behave as recessives. It has not been possible to produce F2 progenies as these hybrids in spite of going through the motions of spawning have failed to produce any progeny. It is possible that they are sterile or that sexual maturity is delayed. Efforts are being made to attempt backcrossing to the parental types.

We shall return later to the question of the genetic control of colour when we consider the effects of the third colour component, guanine, on coloration. Suffice it to say for the present that the guanine-containing cells, the iridophores, are responsible for the silver or white colour seen in variegated fish and their presence confers the metallic sheen of the common (and other metallic) goldfish and renders the scales opaque. In their absence the scales are transparent and this can produce an extended colour range which we shall consider.

The twin-tail mutation

This highly significant mutation in the evolution of the goldfish can with strong reason be considered unique. It has been known for approximately 500 years and nothing comparable has occurred and survived in any other domesticated ornamental fish species.

The genetic analysis of the character has not been easy, largely because it was attempted about 70 years ago by Matsui (1934) when a great many important concepts regarding gene action and phenotype development had not been defined. Smartt and Bundell (1996) re-examined and re-interpreted Matsui's data and came to the following conclusions. Matsui's observations are explicable on the hypothesis that the twin-tail character is controlled by a single incompletely penetrant dominant gene. This resolves the difficulty which Matsui had in interpreting his observations that while twin-tails could produce single-tail progeny, some single-tail segregants from his crosses produced progeny with twin-tails on spawning. The anomaly is explicable if the mutant is considered to be dominant but its expression is subject to strong environmental influence. It is a matter of common observation among the breeders of twin-tail goldfish varieties that spawnings which occur at relatively low temperatures produce high proportions of single-tail progeny. On the other hand spawnings of such stock in the summer produce a very high proportion of twin-tails. Subsequent observation indicates that the temperature at which embryonic development occurs is critical rather than that at which spawning took place. For this reason twin-tail breeders who spawn and rear their fish at ambient temperatures in Britain prefer to spawn their fish in late spring and early summer. If for any reason they wish to raise early broods then they will rear their fish at temperatures of 20–25°C to maximise penetrance of the mutant gene.

Penetrance of mutant genes is concerned with whether or not the mutant phenotype develops in the appropriate genotype. It can be considered as an all-or-nothing situation: either the mutant phenotype develops or it does not. However, when individuals which show the mutant character are examined, the extent to which it is developed can be seen to be variable. At one extreme both the caudal and anal fins can be seen to be duplicated completely; at the other, both these fins may be single. However, fish are produced, in some progenies at least, where development of the anals is completely inhibited and that of the caudal partially or completely so. Smartt and Bundell (1996) suggested that the effect of this mutant gene is one of developmental destabilisation and that through its action a range of developmental outcomes may occur. The development outcome in individual cases is probably the result of the interaction between the mutant gene itself, the background genotype and the environmental conditions prevailing during development, especially in the embryonic phase. The extent to which the mutant phenotype is developed is termed expressivity and in the case of the twin-tail mutation shows an extraordinary range and apparently affects not only the caudal and anal fins but is probably implicated in the genetically produced variation in development of the dorsal fin also. It is significant that lineages which have lost the dorsal fin are all twin-tailed stocks and that partial suppression of dorsal fin development is much more common in these than in single-tail stocks. This implies that the only fins not affected by the mutation overall are the paired fins, pectorals and pelvics, whose evolutionary fate in higher organisms (amphibia, reptiles, birds and mammals) is to be the tetrapod limbs.

The variety and range of effects which can be attributed to the action of this single mutation possibly does not end with the effects on finnage briefly outlined. Hervey and Hems (1948) suggested that the development of the typical short, deep body of the twin-tails owed something to the eponymous mutation, in other words it is a pleiotropic effect. In support of this suggestion, the Wakin, which initially has the appearance of a Common Goldfish with twin-tails when young and immature, at the time it achieves maximum size the body appears to be relatively shorter and deeper than a Common Goldfish of the same size and weight. Experimental verification of this suggestion would be an interesting task to attempt.

The range of fin development which can be produced in twin-tail mutants is certainly worthy of some discussion. The simplest situation is presented by the anal fin which may be duplicated, single or totally absent. In addition there may be asymmetric development of the fins when a pair are produced: one may be fully formed, the other may show less extensive development and may even be vestigial. Other possibilities include incomplete development of a single anal to produce only vestiges of a fin. In other words there appears to be the possibility of a full range of development from total absence to complete duplication. Goldfish fanciers in the West attach a great deal more significance to doubling of the anal fins than those in the East. This may be due to the fact that when viewed from above, as is traditional practice in the East, the anals cannot be seen.

Since the caudals are larger, the range of variation in development is the more readily appreciated. It is interesting to look at the state of duplication of fins in the fish which appear on the de Sauvigny scrolls. Only a couple or so of the fish depicted have fully duplicated caudals. Commonly the fins are united to a varying degree dorsally. The caudals may be united completely in this region, producing a 'tripod-tail' (or tritail); where the union is only along the dorsal margin a 'web-tail' is produced. It is possible also for the dorsal or ventral lobes to be duplicated in part only. A further possibility, when the two caudals are not fused at all, is that their development may be asymmetrical, one caudal may be fully developed while its partner may not be so. The extent of development may range from nearly complete to absent. This is obvious when the caudal fin of an apparent single-tail is not held vertically and may deviate 30° or so from the vertical. When the caudal peduncle is examined closely it may be possible to locate the site where the duplicate fin should have developed.

Apparently perfectly normal single-tail fish can be produced by twin-tail parents, but what has often been overlooked (or ignored!) in the past are those instances where development of the single caudal is incomplete. The range of development can range downward from completely normal development to virtually complete or complete absence. The Meteor might well be an exemplar of the latter condition. I have been informed (J. Linale, personal communication) of occasional Meteor-like individuals occurring in progeny from Veiltail spawnings and I myself have, as noted previously, observed a significant number of segregants in a Fantail Moor progeny which were effectively without a caudal fin, it being reduced to a bristle-like tuft. Tail-less individuals are occasionally reported from the wild in single-tail stocks; the cause of this condition is uncertain, whether due to injury, an accident in development or possibly some genetic cause. The precise cause could only be determined by a progeny test.

It should be noted that although controlled by the same genetic locus, development of the anal and caudal fins shows a measure of autonomy and the state of development of one fin type (the caudal, say) can be combined with a range of states of the other (the anal or *vice versa*). The possible role of the twin-tail mutation in producing the dorsal-less condition was first suggested by Smartt and Bundell (1996) in view of Matsui's apparent failure to identify a specific locus responsible for the condition. It is not uncommon when spawning twin-tail varieties which carry normal dorsal fins to produce occasional progeny individuals with incompletely developed dorsal fins. Sometimes it is the anterior portion which has failed to develop, sometimes the posterior. Selection of individuals with the propensity to produce short dorsal fins and breeding selected individuals, perhaps those in which the anterior portion of the fin failed to develop with those lacking the posterior portion and selecting the progeny for deficiencies in dorsal fin production could perhaps lead to complete suppression of the dorsal fin. It is interesting to note that in the broods of many dorsal-less varieties and strains that individuals occur which have vestiges of a dorsal fin, sometimes only a spine or a few rays and sometimes only a few lumps and bumps where the dorsal fin should be. These observations suggest that the potential to produce the dorsal fin has not been lost but that it is being suppressed. The means by which this suppression has come about is the progressive accumulation of genes which individually have a small depressant effect on dorsal fin development and which ultimately can suppress it completely. This hypothesis is supported by the observation that it is now possible to find lines of dorsal-less varieties which effectively breed true; presumably the genotype for dorsal fin suppression has been fixed. Those lines which continue to produce spikes, vestiges of fins and saw-backs have not yet achieved fixation.

The significance of polygenic (i.e. multifactorial) inheritance in goldfish genetics has only comparatively recently been recognised and we shall come across further examples as our discussion continues. A case in point is the change which can come about in the dominance relationships between mutant alleles and their wild-type alleles at loci where the mutant acquires a selective advantage. We have already seen this in the case of the xanthic mutation;

whereas we have noted xanthic mutations in species other than goldfish are often recessive, that in the goldfish has acquired dominance. A similar situation has developed with regard to the twin-tail mutation, which Smartt and Bundell (1996) suggest is an incompletely penetrant dominant. The cross with the wild-type Crucian carp of red metallic Veiltail × Crucian carp gave all single-tail progeny (100+) with the exception of one individual which showed development of a small duplicated interstitial segment in the ventral lobe of the caudal. As in the case of the xanthic mutation, the influx of 50% wild-type background genotype alleles from the Crucian carp has sufficed to reverse the dominance relationships of the two alleles at the locus producing the twin-tail effect. Dominance evolved apparently by the selection of multiple genes (polygenes) which enhance favoured aspects of the mutant phenotype in the presence of at least one mutant allele.

Fins – size and shape

Fins range in size from those of the Common Goldfish (to all intents and purposes identical with the wild-type progenitor), in which if we take the caudal as exemplar is about one-quarter to one-fifth the body length, to the Veiltail in which length may be as much as twice the length of the admittedly shorter body. There is a broad general proportionality between the size of the fins: a long caudal for example will be generally associated with a high dorsal and relatively long pectorals, pelvics and anals. The variation in fin size as measured by length, say, is continuous; in a progeny segregating for size of fins it is not usually possible to classify them in terms of discrete classes such as long versus short. If size frequencies are plotted they will tend to produce something approximating to the bell-shaped curve of the normal distribution. This is typical of the segregation pattern produced for a character under polygenic control. No fixed segregation ratios (e.g. 3 : 1) are produced. Numerous genes (or genetic loci) are involved, the alleles of which differ according to whether they affect the relevant character positively, negatively or not at all. The effects of these loci are additive, so a long tail can be regarded as having been produced by a genotype which includes a preponderance of alleles having a positive effect while few if any would occur in an individual which had a tail no longer than that of the wild-type.

For a pair of fish to breed true for fin size, appropriate alleles would have to be fixed i.e. homozygous. In fact the longest finned varieties such as the Veiltail commonly do not, implying that all the alleles of positive effect are not fixed. An important feature of genes which control quantitative or metrical characters such as size or length is that they do not show dominance. What is seen is the additive effect of all the relevant genes present. This situation was well exemplified in the F1 progeny of the cross red metallic Veiltail goldfish × Crucian carp. This was quite variable in length of finnage produced although all were within the parental size range, some finnage was only a little larger than that of the Crucian carp parent while the largest was more nearly intermediate between the two parents. Had the parents both been homozygous for the relevant genes they carried, the F1 would have been uniform, that it is not is due to the heterozygosity of the Veiltail parent which characteristically produce variable progeny in Veiltail × Veiltail matings in contrast to the Crucian carp where uniformity is characteristic of its progeny.

Genetics of the shape of fins has been little studied; however, there is much of interest here to the breeder. There is considerable variation in the shape of fins, most particularly the dorsal and the caudal. The other fins vary in size (already noted) and in shape, whether they are pointed or rounded in outline. One aspect on which a tentative conclusion might be drawn is that the bifurcated tail character is dominant to the straight trailing edge of the caudal. This, at any rate, is indicated by the interspecific cross goldfish × Crucian carp in which the F1 have strongly bifurcated caudals.

Two points of particular interest regarding the Veiltail and other broadtail varieties are the shape and size of the dorsal and caudal fins. Size of itself as already noted is a polygenic or quantitative character and this view probably applies equally to the breadth of the tail. This view is consistent with the observations made in Veiltail × Crucian carp hybrids where length and breadth of the F1 caudal were in the intermediate range between the parents. The characteristic straight edge of the caudal could well be recessive with the possibility that perhaps two complementary loci might be involved, for whereas both parents in the cross had fairly straight-edged caudals the F1 caudals were markedly bifurcated. Second only to its magnificent tail, the dorsal fin of the Veiltail is one of its crowning glories. Its height, it is reasonable to conclude, is a quantitative or polygenic character but its distinctive feature, the shape, may not be. There is some variation in the shape of the dorsal in the Veiltail. At one extreme we have a relatively short fin (which may nonetheless have reasonable height) in which the leading and trailing edges are more or less parallel, and at the other extreme the planes of leading and trailing edges make a right angle with each other. The combination of this feature with appropriate length and height produces the unforgettably magnificent sail-like dorsal fin of the Veiltail.

Our understanding of how the qualities we wish to see in the finnage of our fish are produced in the course of development is improved if we refer back to some aspects of embryonic and larval development. Although at the time of hatching the larva has no fins, the die has been cast as far as many aspects of potential fin development are concerned (Figs 1.3a, 1.3b, 1.3c). The form of the caudal peduncle, whether the individual will carry a single or twin tail, is apparent. The determinant of the development of finnage, the fin-fold, is already established. The extent of this is related to what fins actually develop. If there is no fin-fold in the anal region the anal fin will not develop and if it is suppressed in the dorsal region a dorsal-less fish will be produced. Superficially all larvae of the goldfish at hatching look alike but if we know what to look for there are unmistakable indicators of what the future may hold.

Some concluding comments on fin development post-hatching are perhaps appropriate. Fin growth and development in long-finned varieties is progressive and it has been observed consistently that in such varieties fin growth continues during life at what appears to be a disproportionate rate. For this reason precocious development of long finnage, which Innes considered to be a feature of Veiltail runts, is highly undesirable. The 'World's Fair Fish' it should be noted developed its remarkable features over time. Fish which produce impressive finnage early in life are likely to become over-finned with age and not to retain their quality in the longer term. We should perhaps be looking for long-finned varieties to achieve their peak not inside three years but after five years or longer. Then we can hope that this peak of condition and development will be more sustainable.

General coloration

The coloration of fish is rather more complex than that of the higher vertebrates, most notably the mammals, in that three classes of pigment cells and hence pigments are involved. To recapitulate, melanin is present in the melanophores, xanthic pigments are found in the xanthophores, and guanine occurs in the iridophores. The pigments interact to produce the cryptic colour seen in the wild-type goldfish; the attractive coloured variants we commonly see are produced either by loss of one or more type of pigment cell or an increase in their number. The function of the three types of pigment cell (or chromatophores) is related in an interesting way; they all have a common origin from a particular type of stem cell produced by the neural crest of the embryo (Bagnara *et al.* 1979). This implies that they have in common some of their biosynthetic repertoire in addition to the more specialised and specific functions that each type carries out. As a result we can have gene mutations which may affect the function of all three types or which are more specific to only one or perhaps two. We have already seen an example in the depigmenting genes of Kajishima of a mutation which manifestly affects the melanophores and probably in addition the xanthophores. The interrelatedness of the genetic control of development in the three types of pigment cells therefore does not concern only the ultimate interaction between their final products.

In the course of development of the chromatophores they produce within the individual cells bodies (or organelles) which produce melanin, xanthic pigments or reflective plates of guanine. As a general rule only one type of pigment is produced in each cell but in birds some melanophores are reported as also producing guanine. Although the products found within chromatophores are very different it seems that they have much in common in the way they operate and their potential. Development of melanophores has been studied more than that of the two other types. They may vary in size and extent of branching. They are also under the influence of light and may expand or contract in response to light intensity; at low light intensity they contract and at higher intensities they expand producing a darkening of the colour. Their structure is more readily observed than that of xanthophores. As Affleck (1952) has pointed out, although the xanthophores can be shown to have a structure comparable to that of the melanophores, the cells are less robust and without use of appropriate preparation techniques collapse. Affleck identified a carotenoid pigment in xanthophores, found in vesicles within the cells, but in more recent work (Bagnara *et al.* 1979) the more important pigments are thought to be the pteridines. It is common not only in fish but also in amphibia to find these two types of pigment occurring in association.

Colour differences arising from genetic causes have been more intensively studied in the zebra fish than the goldfish (Kelsh *et al.* 1996). Numerous genetic mutations have been identified which cause differences in pigmentation through loss or modification of one or more classes of chromatophore. Further in-depth information on developmental and genetic aspects of chromatophore function can be obtained by consulting the bibliography of the Kelsh paper. The biochemistry of these pigments is a complex topic which can only be considered briefly in the present context.

Affleck (1952) gave a very helpful basic account of the disposition of the various pigment cells in the skin of the goldfish, which aids considerably our understanding of how the colours we observe in fish are determined by the nature and location of the pigment cells (Fig. 1.6). Chemistry is by no means the sole determinant of coloration; it is considerably affected by physical considerations related to the location and the concentration of the pigments themselves as well as the relative densities of the chromatophore cell populations. He considered separately the distribution of the reflective tissue comprising the iridophores and that of the melanophores and xanthophores. The guanine-containing iridophores are confined to two locations. They occur as a backing on the inner surface of the scales which confers the metallic reflective character of scales in which it is present. This layer predominantly develops in the exposed portion of the scale and is not physically a continuous layer. There is a deeper seated more or less continuous layer of iridophores situated towards the inner surface of the dermis external to the muscle and adipose tissues. Normally this layer is not visible unless the scales are transparent. Melanophores may occur in four possible locations:

(1) Immediately below the epidermis on the outer surface of the scales producing the visible coloration of metallic fish
(2) Immediately below the scales
(3) On the surface of the adipose tissue at the level of the inner iridophore layer
(4) In the outer portion of the muscle tissue.

The xanthophores are most obvious in the first two of these locations, while locations (2) and (3) coincide with the iridophores.

In the light of this basic information we can consider how the mutations which affect development and survival of the three types of pigment cells affect the colours and patterns in our fish. We have already considered the apparent paradox that the development of brilliant colour in the goldfish depends on the loss of melanin pigment in the case of the xanthic mutation. Perhaps even more far-reaching is the effect of the complete or partial loss of the guanine-containing iridophores, which makes visible the deeper-seated pigment layers. This opens up possibilities for development of coloration arising from the optical properties of tissues which lie between deep-seated chromatophores and the surface. For this reason it is logical to consider first the effect of reduction or loss of the iridophores on the development of colour.

The effects of iridophores on coloration

Two mutations are known in the goldfish that affect suppression of the iridophores. The first is the 'transparent' mutant of Chen (1928) which is generally regarded as an incomplete dominant (*T/t* locus). This in the homozygote can produce virtually complete suppression of the three types of chromatophores in the skin. In the heterozygote almost complete suppression of iridophore production associated with the scales occurs but their development in the dermal layer is usually quite extensive. This imparts a mother-of-pearl or nacreous sheen characteristic of the Shubunkin and other calico variants. The transparency of the scales means that the deeper-seated chromatophores and the colours they produce are visible. This results in two classes of transparent-scaled variants. In the presence of the depigmenting gene, transparent scaled self-coloured red or variegated fish are produced. Thus, for example, Sarassa Comets may have metallic reflective scales or may not possess highly reflective scales but in which the red and white coloration is combined with the nacreous sheen. In the absence of the depigmenting mutant allele when melanophores are present, colours are possible in addition to the typical black and brown, dependent on their position below the scales, their density and the nature of the refraction effects. The effect produced is a complex mottling. The form of mottling which appeals most to the fancier is when a basic blue background is overlaid with patches of brilliant red or vermilion, with a variable admixture of violet, orange, yellow and even green, the whole of which is in turn overlaid with intense black or brown spots and blotches.

Chen (1928) suggested that mottling was produced by the action of the transparent gene but Matsui (1933) disagreed. The occurrence of mottling is in the first place dependent on the absence of active depigmenting mutant alleles, but in all probability the effects desired by the breeder are determined by a complex of multifactorial modifiers. It should be emphasised that white or silver coloration results from the presence of iridophores and the absence of melanophores and xanthophores. When iridophores are also absent, fish have a pinkish, matt appearance where the visible colour is produced by the haemoglobin of blood in superficial capillary vessels. The effects of the transparent mutant do not necessarily completely suppress the survival of the three chromatophore types. We shall consider later how it is possible to modify and reduce this suppression and the important effects this can have.

Affleck (1952) outlines the terminology which has found favour regarding the phenotypes produced by the transparent gene. Since it is an incomplete dominant, each genotype has a characteristic phenotype, as follows:

Metallic	*tt*	Guanine deposits on the scale produce a reflective body surface, no pigments below the scales are visible.
Nacreous	*Tt*	The great majority of scales are transparent, and the deeper seated iridophores are visible which impart the mother-of-pearl or nacreous sheen. The deeper seated chromatophores are also visible.
Matt	*TT*	Suppression of chromatophores of all three types may be virtually complete. In the virtual absence of any reflective tissue a matt effect is produced.

A second mutation suppressing production of iridophores was encountered by Chen in the course of his genetical analysis of the original transparent scale mutant. He noted that the reduction of iridophore production was apparent in the exposed portion of the scale. Matsui (1934) described this phenotype as 'net-like transparent' and showed that the mutation was recessive. Later (Matsui 1972) he recorded the discovery in England of a similar mutation by Miss Daphne Morris which was called 'mock metallic'. Mr Al Thomma (personal communication) found in feral goldfish populations in Ohio a form which was known locally as the 'bluebelly'. All three forms (net-like, mock metallic and bluebelly) produce scales which may have only a partial backing of guanine. Tippit and Bennett (1965) studied the effect of environmental factors on development of the phenotype and showed that fish raised out of doors approached the wild-type in development of guanine, while fish reared indoors produced many completely transparent scales. One of the names given to this mutant – the bluebelly – derives from the propensity to produce transparent scales in the abdominal region, through which the deep-seated melanin in the body wall is visible as a blue coloration. Mr Dave Mandley has made an experimental cross between a mock metallic (derived from a Shubunkin) and a bluebelly, which produced progeny which showed similar reduced guanine production to that of their parents. Had these two parents carried different recessive mutant genes the progeny would have been normal metallics.

An important difference from the transparent gene (*T*) is that only the production of iridophores is affected; that of the xanthophores and melanophores occurs normally. In the presence of the depigmenting gene the customary colour change occurs; however, the quality of the colour produced is somewhat different. The net-like transparent is also compatible with the melanic mutant (as in the Moor) to give all-black fish. However, Matsui noted that no calico type mottling results from the action of the net-like gene on its own. There is increasing interest in the use of the net-like mutant in the enhancement of colour quality. Miss Morris pioneered the development of the pseudo-matt (a combination of heterozygous transparent and mock metallic) in which calico colours were enhanced through, it was thought, the reduction of guanine. The net-like mutation causes reduction of guanine without affecting other colours. Although the pseudo-matt had enhanced coloration and soft gill plates they had black 'button-eyes' which were not greatly appreciated. There is undoubted scope for genetic manipulation of 'net-like' in combination with other mutations in the enhancement of goldfish colour.

The effects of melanophores on coloration

We have already considered some of the effects of melanophores and the melanin they contain on pigmentation. Both their presence and absence are equally important. Loss of melanin as we have seen is significant in having been the event which produced the Common Goldfish as we know it. However, melanin production can be enhanced and where this has occurred, striking and handsome fish, most notably the Moor, result. In combination with the transparent gene we have the striking blue coloration produced by deeper seated melanophores. An interesting feature of the xanthic mutant is that though it has lost its melanin from the skin, it is not an albino because melanin is still produced in the eyes. True albino goldfish are known which have pink eyes and no melanin, they are often called 'lutinos' because they are commonly yellow in colour but sometimes orange or reddish. In the presence of the depigmenting gene a completely white fish can be produced. Albinos have become available commercially in recent years, and though of considerable intrinsic interest, they do not seem to have a particularly strong appeal.

An intriguing feature of the intense black coloration of the Moor was the fact that there appeared to be a very close linkage between the telescope eye and the melanic factor and that it was very difficult to obtain black strains of fish other than telescopes. This linkage now appears to have been broken as black variants of the Oranda and Ranchu are now commonly seen. Experience in breeding black Lionheads has been rather perplexing in that very few if any black fish have been produced from spawnings. This is in contrast to Moors which breed true. It is possible that there are two different systems of melanic production; that operating in the Moor is relatively insensitive to variation in environmental conditions while that of the Ranchu is affected by them perhaps.

Variegation is a feature of xanthic goldfish as typically expressed in the Sarassa types. In recent years something of a sensation was created by the appearance from the Far East of black and white telescopes which were known to the trade as Magpies or Pandas. These fish are of particular interest in that they have lost their xanthic pigment as well as a substantial part of their melanin. In the Moor we have the situation that development of melanophores is so abundant that melanin totally masks any xanthic pigment which may be present. Interestingly and particularly when the fish grow older, the intensity of the black colour on the belly may lessen and in some fish the colour is bronze while in others it is grey or bluish. The implication of this colour difference is that some of the Moors have lost some or perhaps even all of their xanthophores. As and when melanophores are lost in the production of variegation, two distinct colour patterns result, black and white on the one hand and black and red on the other. The latter I have heard are called 'Pandas' in the Far East, after the Red rather than the Giant Panda. Black and white variegated fish are called 'Magpies' there.

There are two other melanin based colours, brown and blue. Brown, somewhat confusingly referred to in some literature as purple, is probably best exemplified in the Chocolate Oranda. In these fish some particularly attractive shades are developed. Brown colour results from the simultaneous presence of melanophores and xanthophores, both contributing to the ultimate colour. An interesting observation of Mr J. Bundell is that the alevins of brown fish can be distinguished by the colour from metallic alevins of all other colours which at this stage are indistinguishable. There is the possibility here that the brown pigment differs somewhat in its chemistry from the commoner form of melanin. The biochemistry of melanin is notoriously complex and there is little chance of an answer at this time. The size, shape and melanin content (and perhaps composition) of melanophores is variable, as reported by Chen (1928) and other authors. There is the possibility that there is a wide range of genetic effects here which have not as yet been explored and which in present circumstances are unlikely to be studied in depth. Useful information may be forthcoming from reported observations of enthusiast breeders on experimental crosses. Unfortunately little credit would be gained by any institution or professional researcher undertaking this work in this age of biotechnology and genetic modification. The best we can do in the circumstances is to appraise the situation, devise working hypotheses and refine these in the light of experience.

Blue coloration in goldfish is due to melanin and three different intensities of blue colour are produced. The most intense is the blue-black colour produced in the metallic Seibun type (Watanabe 1988) which originated in China; the commoner form is seen in the 'blue' Oranda which is a blue-grey colour. The most uncompromising blue is that of the transparent scaled calicoes which varies in intensity but is not uncommonly a deep forget-me-not blue.

It is apparent from this brief consideration that the role of melanin, contained in the melanophores, in colour determination is complex. We can identify the following variables which affect the outcome.

(1) Presence or loss (partial or complete) in producing self-coloured xanthic mutants or variegated black and white or red and white forms as well as true albinos.

(2) Stability of variegation. In Pandas we have the stabilisation of what is usually the transitional stage in colour change where intensification of melanin development is followed by loss of melanophores by apparent genetically programmed self-destruction of melanophores. It has been suggested that this is triggered by the excessive production of toxic intermediate compounds in melanin synthesis which causes melanophore death. Some genetic means of halting this process apparently occurs. It also appears that the normal sequence of depigmentation can be reversed, as in the Magpies where complete dexanthification is followed by partial demelanisation. In the case of some

Moors and all blue metallics, dexanthification may be achieved without any apparent loss of melanophores.

(3) Number of chromatophores. The black coloration of the Moor is achieved by enhanced production of melanophores, the melanin of which masks the effects of both xanthophores and iridophores when these are present.

(4) Melanin content is a variable which has been noted but not investigated. By analogy with the medaka (*Oryzias latipes*, a toothed carp native to Japan) and mammals in which multiple allelic series occur, where the various alleles produce different concentrations of melanin, it is possible that a similar situation may occur in the goldfish although no evidence for such a series has yet been found.

(5) Variation in melanophore size has been observed by Chen (1928). This is inversely related to the chromatophore number, at least in part: in paler areas the chromatophores are larger than in those which are more deeply pigmented.

(6) Location of pigments is only a significant factor in transparent scaled fish. There is considerable variation in the concentration of melanophores below the level of the scales and these differences are responsible both for the quality of the colour and its intensity.

The effects of xanthophores on coloration

In the xanthophores is the location of the pigments which confer red, orange or yellow coloration to the goldfish. There seems to be a general misapprehension among the generality of goldfish breeders regarding the nature of the responsible pigments. Affleck (1952) mentioned carotenoid pigments and this view has been perpetuated in spite of the fact that another class of pigment, the pteridines, has been identified and associated with the carotenoids. The interest in carotenoids has been sustained by the fact that basically their carbon skeletons cannot be synthesised in the animal body but originate in plants and are passed along the food chain. The belief is current that feeding carotenoids in the diet will achieve colour enhancement in a direct fashion. It is certainly the case that flesh-colour in salmonids is produced by carotenoids ingested in the diet, but this is a very different situation from that in a variegated Sarassa fish where there are distinct pigmented and non-pigmented areas.

The professional ichthyologist is much more concerned with the other class of pigments, the pteridines. These are endogenous pigments, that is produced by the fish itself and therefore subject to change which can be brought about by alteration in the genetic constitution in a way that exogenous pigments would not. This is not to say that changes could not be rung on the ingested carotenoids, but they would be relatively minor.

The commonest xanthic colour seen in goldfish is an orange-red. The quality of this colour is subject to environmental influence. When orange fish spend a summer in a pond under an optimal light regime they may develop an intensely red coloration. (Not for nothing is the common French name for goldfish *poisson rouge*!) This change in colour arises no doubt from an increase in pigment concentration. It is often noticeable that in variegated fish such as the Sarassa the intensity of pigmentation in coloured areas is greater than that in comparable self-coloured fish. There is a possibility that variegation of itself serves to concentrate the pigment in the reduced number of chromatophores. A similar effect is seen with a drop of blood on a slide; before it is spread it is bright red, when more thinly spread it becomes more orange.

An interesting and popular variant of the Common Goldfish is the chrome-yellow which has been established as a true-breeding line. This colour is genetically determined and no amount of feeding astaxanthin or any other carotenoid is likely to change it. The colour change appears to be under simple genetic control but no published account of its nature has as yet appeared. Affleck (1952) suggested, from his own observations that yellow and reddish patches could occur on the same fish, that there was a possibility of both red and yellow pigments being produced. This has not been explored genetically, but it is possible to explain the production of deep, ox-blood red, orange and yellow fish if there were separate genetic control of distinct red and yellow pigments. In deep red fish production of yellow pigment might be suppressed, in the case of the yellows the red pigment would be absent, while the orange fish would produce both. One possible effect of carotenoids in food which is perhaps overlooked is as a source of vitamin A, of which carotenoids are precursors. Enhancement of vitamin uptake could enhance health directly and indirectly improve colour.

Some genetical conclusions on coloration

In summarising our conclusions on the genetical control of coloration we need to consider the effect that mutations have on the three types of pigment cells, the melanophores, xanthophores and iridophores. The major effect appears to be one of the suppression of the development or survival of one or more types of chromatophore. In only one case, that of the Moor, does there appear to be enhancement of chromatophore development. Equally, chemical changes in the pigments themselves do not seem to be of the first importance although it is difficult to avoid the conclusion that the chrome-yellow pigment seen in some metallic goldfish could be chemically different from the commoner orange-red type. The possibility of chemical differences in the melanin produced in black, brown and blue metallics can, at the present time, be neither accepted nor dismissed. As is frequently the case we must bear in mind the possibility that other genes in the overall genotype (the genetic background) may affect the expression of colour mutants and modify the phenotype developed. We need also to review the nature of genetic control: whether a single factor (or locus) is involved or more. If two factors are involved, whether these act in an additive fashion or whether an interaction such as epistasis occurs should be determined.

Bifactorial mutants

The depigmenting mutation, loci Dp1/dp1 and Dp2/dp2

Kajishima (1977) elucidated the genetic control of this basic feature of domesticated goldfish. He showed that two pairs of alleles *Dp1/dp1* and *Dp2/dp2* were involved which were essentially identical in their genetic effects. He suggested that the occurrence of duplicate factors in this instance was a consequence of the polyploid (tetraploid) constitution of the species which could generate such a pattern of tetrasomic inheritance. A regular transmission pattern for these genes is promoted by small chromosome size and consequent low chiasma frequency. There is, however, the possibility of multivalent formation at a perhaps low frequency resulting in occasional production of unexpected progeny types. Tetrasomic inheritance raises the possibility of dosage effects. The allele producing depigmentation is dominant (in contrast to xanthic mutants of other fish species) and this occurs from an age of approximately two months to several years. The timing of the onset of depigmentation, Kajishima suggested, might be determined by the number of active alleles present: this would be most rapid in *Dp1Dp1Dp2Dp2* individuals and least rapid in *Dp1dp1dp2dp2* and *dp1dp1Dp2dp2* genotypes. A high number of active alleles might well be a factor in determining whether both melanophores and xanthophores are lost in sequence or whether only melanophores are affected. It is equally possible that differences in background genotype determine the point at which depigmentation ceases. Loss of melanophores and/or xanthophores may only be partial producing variegated fish, most commonly red and white (silver), red and black and black and white.

Albinism, loci P/p and C/c

Yamamoto (1973) and Kajishima (1977) studied the inheritance of albinism. It was shown that two loci are involved, *P/p* and *C/c*. These show an epistatic interaction. The final segregation produced is 15 pigmented to 1 albino, but studies during embryo development show that initially one-quarter of the progeny are apparently albino but of these three-quarters subsequently develop melanophores in the eyes and skin. Thus a transitory 12 : 3 : 1 segregation ratio ultimately becomes 15 : 1. This arises from the fact that the dominant allele *C* is epistatic to both *P* and *p*, *ppcc* is the persistent albino class while *P-cc* are the transitory albinos. The genetics of this situation is obviously different from the preceding example.

Monofactorial mutants

Blue/brown, locus B/b

The simplest of this class was studied by Chen (1934) and concerns the production of 'blue' metallic fish. These are of a steely-blue or grey colour and are produced by the recessive allele *b* at the *B/b* locus. In homozygous mutants (*bb*) the formation and survival of the xanthophores are suppressed. The heterozygote and homozygous dominant *BB* are brown in colour. On occasions blue fish with a brown patch are seen. These have not been the subject of any published investigation but it is possible that the *B* allele may be incompletely penetrant with variable expressivity. The relationship between this locus and the production of the navy blue coloration of the Seibun has not been explored as yet.

Scale transparency

Three different mutations have been identified and described, the commonest and first to be investigated by Chen (1928) was the *T/t* locus which showed incomplete dominance of the mutant allele, the second, studied by Matsui (1932), was the net-like transparent and the third, reported by Kajishima (1977), is at the *G/g* locus where *gg* produces transparent scales. It is of interest to note the occurrence of this last but it is not of practical significance as yet in any recognised goldfish variety.

The dominant transparent, locus T/t

This mutation has apparently been known for several centuries but it was not until the latter part of the nineteenth century that it came into its own. At this time it occurred in the Sanshoku-demekin or Calico Telescope where it was the most important factor in the production of the multi-coloured, mottled phenotype. The range of colours and shades developed exceeded that in any other variety at the time. By dint of crossing and backcrossing, the calico feature was incorporated in a range of varieties, most notably the Veiltail, and contributed to the development of a novel variety, the Shubunkin, in about 1900.

Since the transparent mutation shows incomplete dominance then each of the three possible genotypes *TT, Tt* and *tt* produces a different phenotype. These have been termed matt (*TT*), nacreous (*Tt*) and metallic (*tt*). They are named on account of the effects each genotype has on the production of guanine. The matt homozygote produces very little guanine and as a result the scales are completely transparent, neither is guanine produced below the level of the scales. As a result of the virtual absence of guanine, there is little or no light reflection from the body, which is semi-transparent. In the past these have been called crystal fish because it is possible to see for example the orange colour of the ovaries through the body wall of females. The nacreous heterozygote produces predominantly transparent scales with occasional reflective scales having a guanine backing; there is, however, a continuous layer of guanine below the level of the scales which imparts the mother-of-pearl sheen. In the metallic homozygote all scales have the reflective guanine backing and although there is a deeper seated continuous guanine layer this cannot be seen. The effect of the mutation is seen not only in the iridophores but also in the melanophores and xanthophores which are suppressed in the matt homozygote but the development of which is not inhibited in the heterozygote. This means that the deeper seated chromatophores are visible and, as a result of various optical effects, a range of often very attractive colours. In the metallic homozygote only the chromatophores above the scales are visible and these alone are responsible for the visible colour.

It has been observed by Chen, Matsui and Affleck that the extent to which the transparent mutation suppresses chromatophore development shows variation. In the mutant homozygote development of dark dots and blue coloration can be produced by development of some melanophores, orange or reddish patches by xanthophores and even reflective tissue on gills, scales and beneath the scales. By selection and selective breeding chromatophore survival can be greatly enhanced, so much so that highly coloured homozygotes with, in effect, calico coloration, can be produced. Calico fish, which are highly popular, are not, because of their heterozygosity, true-breeding and so production of highly-coloured homozygotes resembling the calico homozygote would have great appeal to breeders. It is possible by selection to achieve the opposite effect, that is to say produce a heterozygote which looks like the typical matt virtually devoid of guanine, melanin and xanthic pigments.

These interesting effects are the outcome of the interaction between the mutant gene and the other genes of the genome. Where this background genotype is not fixed, interactions producing a wide range of effects are possible, by amelioration or suppression of the mutant's effect on the phenotype or intensification or enhancement of them. This example provides a graphic illustration of the power of selection.

The net-like transparent, locus N/n

This mutation affects only the development of iridophores. The net-like transparency is produced by the presence of iridophores only in part of the visible portion of the scale. Alternative names for the mutant are mock-metallic and bluebelly. The latter name is given on account of the lower iridophore density on the scales of the abdomen and the visibility through the body wall of deep-seated melanin lining the body cavity. The name mock-metallic was given because of the variation which can occur in different environments in the development of the iridophores. Under high natural light intensities this may be close to that in normal metallic scales, under low light intensities indoors iridophore development may be sparse. A characteristic feature of the mutant is the customary absence of iridophores on the operculum, giving soft gill plates.

The chief interest of this mutation has been its use in combination with the dominant transparent gene to produce the 'pseudo-matt'. This is a calico which is also homozygous for the net-like gene. It was first produced by Miss Daphne Morris. The colour intensity developed could be remarkable, this was combined with soft gill covers and black button-eyes (i.e. lacking a reflective iris). The latter feature was regarded as undesirable by many breeders and true pseudo-matts are not easy to find at the present time.

The recessive transparent, locus G/g

Kajishima (1977) isolated this mutation and carried out the appropriate genetic analysis. He demonstrated that it was a single locus effect and in the homozygote (*gg*) production of guanine was suppressed. No effect was observed on other chromatophores. He gave no indication as to whether this mutant could produce the multicoloured, mottled calico effect that the dominant transparent mutation produces. It is significant in this context to note that in the net-like transparent mutant there is no tendency to produce calico phenotypes on its own. These considerations lead us to reconsider Chen's (1928) opinion that the mottled calico effect was a product of the action of his dominant transparent mutation. This idea was dismissed by Matsui (1934) but the fact that the calico phenotype cannot be reproduced in the absence of the *T* allele does suggest that this should be considered as a possible pleiotropic effect of the *T/t* locus. What we do know at the present time is that the calico phenotype requires the presence of the *T* allele and the absence of active depigmenting alleles (i.e. a *dp1,dp1,dp2,dp2*) genotype.

Eye mutants

Telescope eye

The most common and most popular eye variant in the goldfish is the 'telescope' eye also termed 'globe eye' in the parlance of the Goldfish Society of Great Britain. However, the name 'telescope' has a history of well over a century and is used and understood internationally (the term is certainly used in French and German). The character was well established in the eighteenth century and several examples figure on the de Sauvigny scrolls. Matsui (1934) carried out a successful genetic analysis of the mutant and found that the condition was recessive; he assigned symbols *D/d* to the single locus he had identified as controlling development of this feature, the homozygous recessive *dd* being the only genotype to express it. Interestingly Matsui carried out a cross with the wild goldfish (Funa) in which he obtained a 15 : 1

rather than a 3 : 1 segregation. This he ascribed to the presence of an inhibitory factor A with alleles *A/a*, only in the presence of the homozygous recessive *aa* genotype could the mutant phenotype develop. However, following Kajishima's (1977) lead and that of Ohno (1970) a more plausible hypothesis can be developed.

Ohno (1970) who considered the role of polyploidy in evolution has assembled the ideas largely developed by plant cytogeneticists of the consequences of gene duplication on evolution. The presence of additional gene copies slows down response to selection; in the hypothetical situation of the homozygous recessive at a single locus acquiring a selective advantage in progeny of an F1 hybrid (*Aa*) one-quarter will be at a selective advantage. When we have duplicate genes involved say *Aa/Aa*, only one-sixteenth will be positively favoured. This obviously slows down response to selection. One way of improving responsivity is to restore the disomic situation. This can come about by gene silencing through chemical change at the DNA level or deletion by chromosome structural change of the duplicate locus. Some form of gene silencing could explain the differences Matsui observed in his two sets of crosses.

The Celestial

No genetic analysis of the difference between telescope eyes and those of the Celestial has as yet been published. It could well be that this difference is controlled by polygenic factors, which affect the physical size of the eye itself (larger in the Celestial) and determine the rotation sequence where the eyes first face forward and then turn upward.

The Bubble-eye (Water Bubble-eye)

The genetic control of this feature has not been explained in detail. Judging by the considerable variation in the sizes of the fluid-filled vesicles one might suspect polygenic control. It is quite possible that analysis has been attempted, perhaps on the expectation of identifying a single locus effect, but that the experimenter was unable to interpret the results.

Hypertrophies of the head

These are two in number, the so-called narial bouquets or pompons and the characteristic hoods of the Lionhead and Oranda.

Narial bouquets

These are hypertrophies of the nasal septum and resemble nothing so much as spherical powder puffs. It is often apparent that some individuals in segregating progenies show enlargement of these to a small extent. It is possible that if such individuals were bred further they might respond to selection for increased size and a desirable conformation. The genetic control appears to be polygenic in nature and determines in all probability their size and texture. The size should be large within reason and the texture should be dense; loose straggling pompons should be selected against rigorously. At the present time they are added as optional extras to varieties such as the Celestial and they seem to present no particular problem in a breeding programme.

Hoods

The inheritance of the hood in Lionheads and Orandas was studied by Matsui (1934) but he was unable to develop any acceptable genetic hypothesis. His major difficulty was that development of the hood was in his view greatly influenced by diet and environmental conditions. An important feature in selection is to pay special attention to the conformation of the skull, which should be broad and blunt. This provides the best base for subsequent hood development.

Matsui was probably seeking to identify one or more major genes which controlled hood growth and development. From personal observation it seems that often in older and mature fish varieties there is the suggestion of rudimentary hood growth. The magnificent Veiltail which appears as the frontispiece in Innes' *Goldfish Varieties* shows at least a suggestion of such a growth. Such could form the basis of selection for increased size of the growth itself and precocity in development. Some individuals in an F1 progeny from a cross Veiltail × Crucian carp showed development of modest hoods (at least as well developed as those on many Blue Orandas!) although neither parent (both mature individuals) showed any suggestion of hood development. Such fortuitous events could well be the starting point of selection programmes. These would be very much on the lines that Charles Darwin suggested for evolutionary progress of the progressive accumulation of small differences. At the present time we might also view it as a progressive accumulation of genes of individually small effect.

When one examines the variation of hood form, colour, size and texture, its range is enormously impressive, as an examination of Man Shek-hay's book *Goldfish in Hong Kong* reveals. It is a measure of what has been achieved in the twentieth century by persistent, sustained selection by dedicated breeders.

Scale types

The transparent scale, already considered, is not a distinct type of scale as it differs only from metallic scales in the absence of iridophores. There are three distinct scale types which have been recorded in goldfish, only one of which is at all common, the Pearlscale; the others (the Hammered scale and the Mirror scale) are seen only very rarely.

The Pearlscale

The distinctive feature of this scale is that in the centre of the exposed portion of the scale there is a roughly circular thickened area; with a backing of guanine and the concomitant mother-of-pearl sheen, it is well named. The variety named after this feature has a very characteristic body form, being almost spherical. In appearance it looks dropsical but is perfectly healthy;

remarkably in the standard Pearlscale there is a perfect correlation between the dropsical appearance and the presence of pearlscales. The pearlscale character has been transferred by crossing to other varieties as indicated by Teichfischer (1996), most notably the Oranda to produce the Hamanishiki. The causation of the bloated appearance of the pearlscale could be a pleiotropic effect of this mutant. Although genetic control of the character does not appear to have been analysed and published in an accessible account, it has proven to be freely transferable to other varieties by conventional crossing. It is probably a simple recessive.

The Hammerscale

This variant which has been described in publications of the GSGB by Mr M. D. Cluse. It appears to be the reverse of the pearlscale in having a concave surface rather than convex. It has not been reported in recent years but anecdotal evidence suggests it is a simple recessive.

The Mirrorscale (Fig. 7.1)

I have seen a single example of a Common Goldfish with mirror scales, the property of Mr Gordon King, a member of the GSGB. Subsequently reports were published by Mr Merlin Cunliffe, also a Society member, of Comets in Australia with similar scales. Presumably Mirrorscale goldfish have been produced from crosses with Mirror carp (*Cyprinus carpio*), in which the feature is produced by a recessive lethal gene (*N/n*). On mating Mirror carp a 2 : 1 segregation is produced because the *NN* homozygote is unviable. Possibly there would be a similar outcome from comparable matings with goldfish.

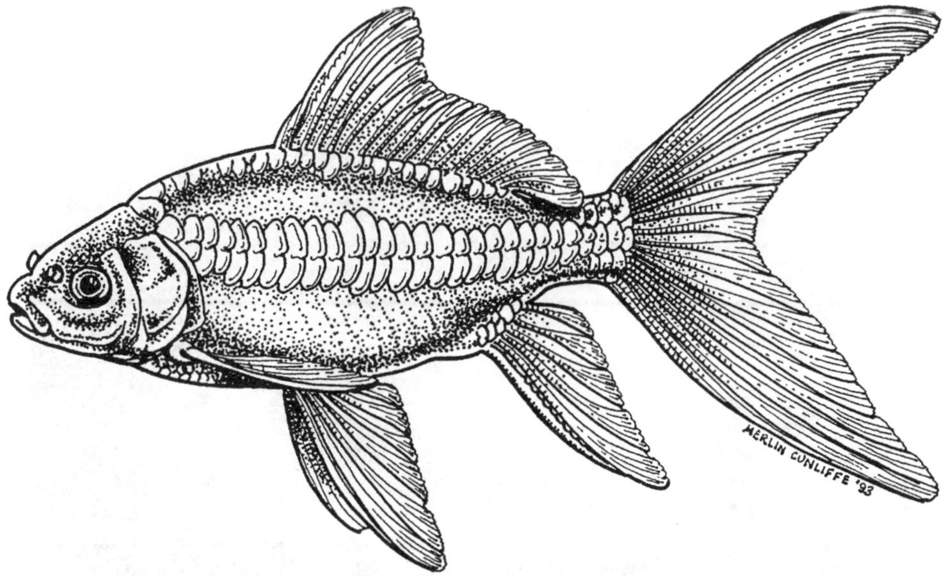

Fig. 7.1 Mirrorscale Comet.

The out-turned operculum

This feature (Fig. 7.2), greatly appreciated by the Chinese, tends to be regarded as a deformity in the West. Teichfischer (1996) considered the range of goldfish variants in which this character had been incorporated. Viewed from above the character is customarily seen as an outward flaring of the operculum with a forward curl. My own experience suggests that actual expression is much more variable and while a stable expression is possible this may only be achieved by dint of sustained selection. It appears to be recessive and comes to expression

(a)

(b)

Fig. 7.2 (a) Veiltail × Crucian carp F1 hybrid showing extreme expression of the out-turned operculum phenotype, with virtually complete suppression of operculum development. (b) Shubunkin × Crucian carp hybrid, showing typical expression of the out-turned operculum phenotype. The operculum is curled, partially exposing the gills. (Lockyer.)

most often when some inbreeding has been practised. I am not aware of any formal genetic studies having been undertaken or published. I produced progeny with abnormal opercula in crosses involving London Shubunkins (a strain from the late Mr W. Leach) and red metallic Veiltails as goldfish parents on the one hand and English Crucian carp on the other. Both the London Shubunkin and the Crucian carp when inbred produced progeny with affected gill plates. The status of the female Veiltail was uncertain. However, reciprocal F1 progenies of Crucian carp × London Shubunkin and Veiltail × Crucian showed high frequencies of abnormal opercula. These varied in the nature and extent of the abnormalities; in the more extreme cases little development of the opercula had occurred, leaving gill filaments more or less completely exposed.

This pattern of development has features reminiscent of the observed range of development found in the case of the twin-tail character. A fish with an apparently normal operculum on one side might show a counterpart on the other with development of anything ranging from fully developed but out-turned to virtual failure of the operculum to develop. The occurrence of such a range of developmental outcomes is indicative of variable penetrance and expressivity. Further investigation could well yield interesting and informative results on factors influencing development. In spite of such deformities of the operculum, affected fish are remarkably tolerant of this pathological condition.

Body conformation

The progeny of many short-bodied twin-tailed goldfish will show variation in the depth of body development in the progeny. This variation is continuous in nature and the segregation is not into recognisably distinct classes, which suggests it is polygenic in nature. An interesting segregation was shown in the Veiltail × Crucian carp cross previously mentioned, in which variation in body depth, although intermediate between the parental extremes, approached both parental states quite closely. Interestingly this was not the case with length and breadth of finnage, development of which was closer to the median, perhaps more loci being involved in the latter instance.

Evolutionary genetics of goldfish – some conclusions

Our perspective of goldfish evolution in genetic terms is based largely on the work of Chen (1925 *et seq.*), Matsui (1934 *et seq.*) and Kajishima (1960 *et seq.*). Through this work we now understand better the genetical events which conditioned the initial domestication as an ornamental fish, namely the xanthic mutation and the second major genetical, event the establishment of the twin-tail mutation. Curiously it was the prime mutation which took the longest to understand, which was the achievement of Kajishima (1977). Matsui's work (1934) provided the basis for understanding the twin-tail situation, the problems with which were not finally resolved until comparatively recently (Smartt & Bundell 1996).

Chen and Matsui both had identified mutations which were transmitted in a straightforward Mendelian fashion (the transparent *T/t*, brown versus blue *B/b*, telescope eye *D/d* (and *A/a*). There were other features of importance, including the two fundamental mutations which were not amenable to such straightforward explanations. There was considerable frustration in those goldfish fanciers versed in genetics that the straightforward, uncomplicated genetical analyses which had been published in Schröder (1976) were not matched in the goldfish. It is of interest that in Schröder's excellent book no genetical analyses of characters in either goldfish or koi are mentioned. Schröder apparently knew enough to let well alone! Even Chen (1956) had apparently (perhaps under political pressure) abandoned faith in a Mendelian explanation of goldfish heredity. It was an unfortunate circumstance that when the interest in and support for genetic studies in the goldfish was at its highest that the science of genetics itself had not yet achieved the level of sophistication which would have permitted adequate explanation of the data collected by the extremely able and committed investigators mentioned. In recent times Kajishima has been the only major author with the interest and resources to carry on in the tradition of Chen and Matsui. At the present time there seems little prospect of generating interest in what is perceived as old-fashioned genetics. We therefore have to attempt to re-interpret or interpret for the first time data and information from the golden age of goldfish research. In practice this means that we have to content ourselves with working hypotheses which cannot for logistical reasons be subjected to rigorous scientific tests but which we can develop and refine in the light of experience.

It is appropriate at this juncture to review the advances in genetic understanding which have helped us to understand goldfish genetics better. Perhaps the major problem was the failure to appreciate the fact that the goldfish is a polyploid. Not only is it so but it reproduces by normal sexual processes. This possibility had been discounted by the leading animal cytogeneticists as late as the 1950s and 1960s on the grounds that an *XY* chromosome system of sex determination was incompatible with polyploidy, because unbalanced *XY* chromosome complements would produce intersexes. This view ignored the situation in some butterfly species where the *Y* chromosome behaves as a sex-determining supergene and in the presence of a single *Y* the individual is male regardless of the number of *X* chromosomes. This removes at a stroke the intersex difficulty.

Polyploidy throws up other problems in the tetraploid goldfish; many of the mutants which goldfish fanciers have selected for in its history have been initially recessives. These in the tetraploid or tetrasomic situation would have to achieve homozygosity before the character could be expressed. Responsiveness to selection in the goldfish has been achieved in two ways. The first is through alteration in dominance relationships between mutant and wild-type alleles which Fisher (1930) has shown occurred in the domestic fowl. This is supported in the goldfish when the xanthic (*Dp*) genes are dominant whereas in many other fish species they are recessive. This change has occurred in a time frame of over 1000 years. The second way in which this objective can be achieved is by gene silencing, which could account for the difference in segregation ratios obtained from crosses of telescope-eyed goldfish with domesticated varieties on the one hand and the wild-type goldfish on the other. This difference is explicable on the basis of the silencing of the redundant locus, converting tetrasomic to disomic inheritance.

The genetic control of a number of characters studied by Matsui (1934) were not readily explicable on the basis of mutational changes in major genes (oligogenes) and we have considered the possibility that these are controlled polygenically, examples noted include the hoods of Lionheads and Orandas and the narial bouquets of Pompons. It seems that the polygenic effects may be highly significant in changing dominance relations between alleles at the same locus, and in modifying the phenotypes produced by the action of mutant genes. When indeterminate numbers of genes affect major gene action in this way these can be called collectively background genotype effects. These effects are ever present and explain why constant phenotypes do not necessarily result from a specific genotype. However, the background genotype when it is effectively fixed by selection can serve to stabilise desirable phenotypes. This process may take a long time. It is a matter of common observation in the Ranchu among dorsal-less varieties that progeny individuals carrying spikes, rudimentary fins and lumpy backs are very much less frequent now than say 50 years ago.

Looking ahead it seems that further evolutionary change in enhancing the quality of goldfish will come about by effective manipulation of the background genotype. This could also improve the yield of desirable phenotypes and enhance efficiency of the use of the breeder's time and other resources.

Evolution of the goldfish – a genealogy

It is not surprising that professional biologists studying the goldfish have attempted to devise systematic schemes or genealogies as a means of summarising the evolutionary history of the goldfish. Matsui has presented his ideas in several works since the 1930s and these have

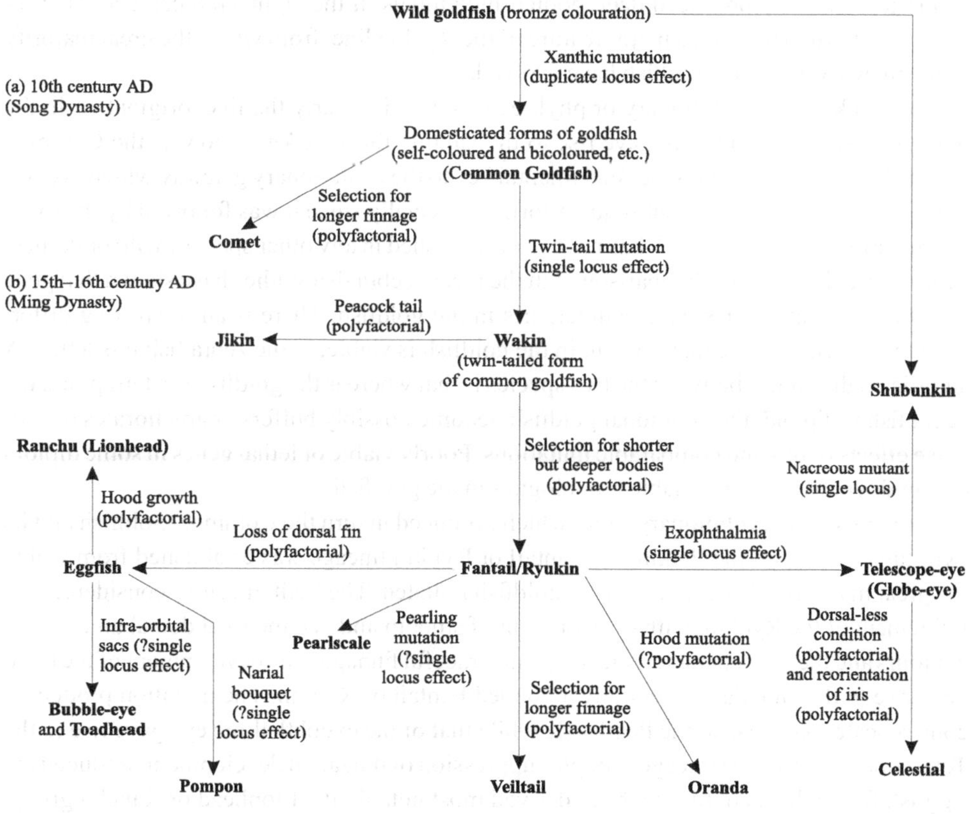

Fig. 7.3 Evolution of the domesticated goldfish, *Carassius auratus*.

been taken up by subsequent authors such as Teichfischer (1994). The scheme presented here (Fig. 7.3) is essentially an updating which incorporates more recent genetic findings and perceptions. Only the major and more significant developments are specifically cited. For example the basic xanthic mutation (a duplicate locus effect) which initially produced orange-red individuals it is thought was followed by others which appreciably extended the range of colour variants, which are not detailed. In this scheme are incorporated variants which are produced on the one hand by single or relatively few genes (oligogenes) and those which are produced by an indeterminate number of genes (polygenes). In addition there are others whose genetic determination has not been adequately studied but whose behaviour indicates the probable nature of the genetic control operating.

While some features such as the establishment of the twin-tail character appear to have been unique events, there are others which could well have arisen more than once and independently in different lineages. Following Matsui's conclusions on the origin of the Lionhead and Oranda, it is reasonable to conclude that the hood could have originated more than once and that perhaps it originated independently in the Lionhead and Oranda lineages. Suppression of dorsal fin development could have occurred independently in the Ranchu and Celestial lineages. Individuals with unusually short dorsal fins arise from time to time in different goldfish stocks which by crossing of selected individuals and further selection could eventually produce dorsal-less individuals. Variation in length of fins within strains could provide the basis for selection of long-finned types. Body depth shows response to selection which surprisingly rapidly produced deep-bodied individuals in the eighteenth century. This response seems to have been a major feature of the Ryukin line, from which the great majority of the more exotic fancy goldfish have evolved.

The trunk of the evolutionary or phylogenetic tree is clearly the line, originating in the wild goldfish, which initially gave rise to the xanthic form we know now as the Common Goldfish. This mutation was the fundamental key to the evolutionary gateway which eventually produced the wealth of subsequent forms, especially since it was followed by the twin-tail mutation, the like of which has not been established in any other species, wild or domesticated, in a viable form. Comparison with the recent zebra fish studies have shown that gene mutations in that species have counterparts in the goldfish. There is an interesting difference in that whereas the matt mutant in the goldfish is viable, in the zebra fish it is lethal. A significant difference between the two species is that whereas the goldfish is a tetraploid, the zebra fish is diploid; the additional goldfish genome possibly buffers or ameliorates the adverse effects of this and comparable mutations. Poorly viable or lethal genes in some diploid cyprinids could well have viable homologues in the goldfish.

From this basic evolutionary trunk which produced in turn the Common Goldfish and its twin-tail counterpart the Wakin, the Fantail or Ryukin lineage was established from which the great majority of twin-tailed fancy goldfish radiated. The Veiltail can be considered as a refinement of the Ryukin, with development of smooth and symmetrical dorsal and ventral contours and the very characteristic, long and graceful finnage. The development of the hood gave rise to the Oranda – essentially a hooded Fantail or Ryukin. The mutation producing convex scales gave rise to the Pearlscale while that of the exophthalmic eye gave rise to the Telescope or Globe-eye lineage. Simple suppression of dorsal fin development produced the Eggfish from which could have been derived most notably the Lionhead or Ranchu group, the Pompon, the Bubble-eye and Celestial. It is a moot point whether all five dorsal-less types constitute a single lineage or not. It is interesting that postwar Celestials, Bubble-eyes and Pompons shared a distinctive body form much more elongated than that of the Eggfish, which suggests the possibility that these three varieties have a close affinity. The alternative is that the particular characteristics of these varieties arose independently and were transferred to a common stock by hybridisation, backcrossing and selection. This question is similar to that of the origin of the hood in the Oranda and the Lionhead. At the present time it is not possible to regard either the single origin of the hood or its independent origin as proven beyond doubt, neither is ruled out in this instance nor is it possible to decide the same issue with respect to the dorsal-less feature. The origin of the Shubunkin by hybridisation of the Calico Telescope with a single-tail goldfish is a matter of record and so the potential role of hybridisation in goldfish variety development cannot be discounted. There is the further possibility that a variety such as the Oranda could have arisen both independently and by hybridisation with the Lionhead.

While the bulk of fancy goldfish variety development took place from the Fantail/Ryukin lineage, there are two significant single-tail varieties that must be noted: the Comet and the Shubunkin. The Shubunkin, as already stated, arose from hybridisation and the Comet also could have arisen from a cross between a common goldfish and a long-finned twin-tail with selection for a long-finned, slim bodied segregant. Significant diversification of the Comet has been in development of colour variants, self-coloured red-orange, the variegated Sarassa and the red-capped Tancho with a silver body (*cf.* Tancho Koi). The diversification of the Shubunkin has been more interesting, resulting as we have already seen in the establishment of three distinct lineages, the original (Japanese/American), the London and the Bristol which have become so popular in the West that separate show standards have been produced for each type. Whereas omnibus classes embracing all dorsal-less varieties may be specified at some shows, the lumping together of all Shubunkin types in a single class would be very much the exception.

The Jikin can be regarded as an outlier of the twin-tail division which is clearly derived from the Wakin. It has diverged from the latter in the development of its distinctive peacock tail – kujyaku-wo – and its striking colour pattern of deep red fins and predominantly silver body with limited red coloration permitted on the lower abdomen, mouth and head. This beautiful fish is difficult to maintain at the highest level of quality; the tendencies which the breeder has to contend with are loss of colour in the fins and the tendency of colour to develop more extensively on the body than is considered desirable. The effect of this is to lose the sharp colour contrast between body and fins.

The diversification we have summarised here has arisen from two main types of genetic change or mutation, those which have obvious major effects and those with individually small effect but which collectively can and have brought about very significant change and development. It is very important to bear in mind the fact that all relevant mutations affecting characters such as the twin-tail (oligogenic) and the hood (polygenic) do not act in isolation. They react to and interact with other elements of the genotype, in other words they act in concert with the background genotype. The question of the background genotype has only seriously been discussed previously in Smartt and Bundell (1996), yet there is hardly an aspect of the fancy goldfish phenotype which is not in some way affected by the background genotype. This can be illustrated by a brief consideration of selected examples. The dominance of the xanthic mutation has in all probability been established by the selection of a background genotype favouring dominant expression. The twin-tail phenotype in its most desirable manifestation is likely to have been favoured by the selection of a genetic background promoting this development. In the case of a character such as the hood, which appears to be polygenic rather than oligogenic in its control, background effects seem to be responsible for the range of expression in size, shape and texture which are now apparent. Perhaps the most striking demonstration of background genotype effects is to be seen in the amelioration of the effects of the transparent scale mutation on the development of guanine deposition on the scales and other chromatophores. This change established by selection of 'coloured matts' produced a movement in dominance relationships of the mutant allele from incomplete dominance towards recessiveness. The practical consequence of this change is that it makes development of a true-breeding Shubunkin a distinct possibility.

The recent production of a fascinating array of attractive and interesting variants by the Chinese owes a great deal to the skilful, if unconscious, manipulation of the background genotype, most notably in the Telescope-eye and Oranda lineages. This, together with the achievement of dominance modification in the Shubunkin lineage, shows that innovative possibilities in the diversification and further development of goldfish varieties are by no means exhausted. This theme will be developed in more detail in the next chapter.

Chapter 8
Genetics and Goldfish Improvement

In essence goldfish improvement amounts to genetic manipulation of goldfish populations which have been produced by breeders. The application of genetical principles should on the one hand be an aid to enhancing results and on the other to the avoidance of disasters which can overtake misguided ventures.

Perhaps if we consider a hypothetical case we can come to appreciate what is involved. Let us suppose that a single outstanding fish comes into the possession of a breeder, which is considered to be the basis for the development of a new superior lineage of an established variety or a completely new variety. The first step is to produce hybrids between the selected individual and another which is as close to it as possible. If it is feasible several such crosses should be made to establish as broad as possible a genetic base. Two different procedures can then be followed. The more important is to backcross selected F1 individuals most closely resembling the élite parent to produce a first backcross generation (BC1F1) and also to produce an F2. The latter should provide some indication of the magnitude of the task of producing for the future a breeding population similar to the élite parent. Perhaps a few promising or interesting individuals could be retained. The backcross F1 (BC1F1) can then be taken and selected individuals again backcrossed to the élite parent and also an F2 produced from the BC1F1 to produce the BC1F2. The BC2F1 and the BC1F2 should be compared and both progenies selected and the process repeated. Backcrossing and F2 production should be continued for as long as possible. It is unlikely that more than three or four backcrosses could be produced, but in the process a useful breeding population should have been established.

The purpose of producing both backcrosses and F2s from backcrosses is to monitor the effects of outbreeding. Since the goldfish is a polyploid and has four genomes rather than two, it is buffered to an extent against deleterious effects of inbreeding. This notwithstanding some excellent strains of goldfish have died out or degenerated irreversibly as a result of inbreeding. The results of inbreeding may be morphological (skeletal abnormalities, kinks or unusual curvatures of the spine, irregularities in the finnage, peculiarities of body shape, etc) or they may be less obvious and insidious. Reproductive capacity may be compromised with the result that eventually a line may fail to reproduce; individuals may be viable but effectively sterile. Both these outcomes can be avoided if an effort is made to maintain individuals of earlier generations which are of known fertility on the one hand and free of morphological irregularities or deformities on the other. Another successful method of avoiding excessive inbreeding is careful selection of brood stock so that spawning parents are not too similar in their morphology or coloration. It may be decided for example not to select a pair where each has similar colour and patterning, nor with very long finnage, deep bodies, etc. Selection of parents with divergent morphology produces a varying progeny from which suitable show specimens may be selected without sacrificing genetic variability.

Inbreeding is a two-edged sword: without a measure of it we would not be able to establish consistent standards in recognisable varieties, but carried to excess can produce sterile or deformed stocks. Why does this come about? In maintaining a goldfish variety it is desirable to fix the distinctive characters genetically. As long as the desired phenotype is not produced by incomplete dominants in the heterozygous state this is theoretically possible. However, there are possibilities of complications arising when desirable alleles on a particular chromosome are linked to deleterious recessive alleles. When the desirable character is fixed by selection, so is the undesirable one. If this is repeated over a wide number of characters then the outcome of fixed undesirable genes could be very detrimental, adversely affecting viability, fertility and morphology, producing at worst poorly viable, sterile and deformed progeny. A useful strategy to circumvent this problem is to practise a strategy of recurrent selection. This entails the fixation of as many desirable alleles as possible in a range of potential parent lines. In such a range of lines it is to be hoped that if any undesirable alleles have been fixed they will be different. Suitable crossings would supply the opportunity for genetic recombination to occur which, followed by selection, could in each cycle tend to fix more of the desirable alleles, if selection for viability, fertility and desirable morphology continued. Continuing cycles of selection and encouragement of genetic recombination could improve the genetic constitution substantially and reduce the frequency of undesirable genes.

Overcoming the problem of 'linkage drag', that is the unwitting selection of detrimental genes linked to the desirable, is a very live issue for the breeder, whose best line of defence is to maintain at least two and preferably more sister lines of each variety to avoid irreversible loss of vigour (inbreeding depression), loss of fertility and a high incidence of deformities.

In recent times concern has been expressed regarding the question of genetic erosion, especially in the context of commercial animal and plant breeding. In such enterprises there is set up a blueprint or 'ideotype' which incorporates all the desirable and essential features to be expressed in the final product. If we have numerous breeders all selecting towards the same ideotype and perforce using similar basic parental material, we will produce genetic convergence and genetic variability will be lost. If circumstances change or the ideotype is modified the breeder may well find that his pool of genetic variability is so depleted that it is not possible to obtain any response to selection unless the gene pool can be replenished.

The goldfish breeder is essentially in the same situation. His ideotype is the standard and by continued selection he is narrowing the genetic base, which as we have seen, has both good and bad consequences in the fixation of desirable features on the one hand and undesirable ones and inbreeding depression on the other. If all breeders were to follow precisely the same practice we would inevitably lose viability and fertility in our breeding stocks and we would run a very serious risk of losing varieties. Perhaps even more importantly, if we were to avoid the inbreeding depression scenario, we would be most unlikely to generate new and interesting variants of the kind that have emerged from China in recent decades. The problem arises from the fact that, in the West certainly, many breeders (especially amateurs) operate on a very small scale and their selection procedures are highly focused and they must single-mindedly eliminate what does not conform to their taste (or standard). Limited space and resources make this inevitable. Commercial fish farmers producing for the market, who operate within very tight margins, can rarely afford the luxury of putting time and effort into the development of new variants which may not necessarily be a commercial success. Unless the proprietor is prepared to undertake this kind of development as a hobby, then we are fated to be stuck in a rut.

In comparing notes with breeders over the years, it frequently happens that occasionally aberrant individuals arise which appear to be 'mutants'. I have had occasion to mention a spawning of Fantail Moors which produced butterfly tails, individuals with little or nothing in the way of caudal fins, and with eyes of reduced (rather than larger) size with respect to the normal. The cause of this variation is not easy to explain without extensive experimentation, which alas nobody is prepared to do these days. However, we should if at all possible and if our curiosity is aroused, follow the good scientific dictum of 'cherishing our exceptions'. My experience is by no means exceptional. Mr Merlin Cunliffe has mentioned in correspondence a number of aberrant forms produced in spawnings of a range of varieties, for example, the production of a Moor with upturned Celestial type eyes. As Mr Cunliffe has never kept Celestials, this spontaneous occurrence is consistent with the view advanced earlier that gene interactions are very important in the development of characters with appeal to breeders (Figs 8.1, 8.2, 8.3). These could arise by mutations floating in the population coming to expression, a background genotype effect or by a small structural mutation affecting a chromosome, inducing a position effect.

Expression of such extraordinary features may be one-sided. Mr Cunliffe observed an Oranda with a water-bubble eye on one side only. In a Ryukin progeny he observed an individual with a Tosakin-like conformation of the caudal which, however, was divided and not a web-tail. In the progeny from Crucian carp × Veiltail progenies there were individuals which showed development of hoods and somewhat enlarged narial flaps. In the crosses with the Shubunkin one individual developed telescope eyes. This is an interesting throwback to the original parentage of the Shubunkin, the Sanshoku-demekin or Calico Telescope. A similar throwback was reported by Mrs Pam Whittington when she noted unexpected tripod-tails in Shubunkin progenies. Possibly some cytogenetic event such as a small chromosome structural change has in these instances removed inhibition of expression of the telescope eye and twin-tail characters of the Calico Telescope parent.

Manipulation of dominance relationships between alleles at a genetic locus

It is clear from what has been considered so far that dominance relationships between alleles are not fixed. We have noted already the difference in dominance of the xanthic and twin-tail mutations *vis à vis* their wild-type alleles in domesticated goldfish and Crucian carp hybrids. Recently there has been interest in the behaviour of the dominant transparent mutation (Smartt 1997) in view of the fact that two novel goldfish varieties, the Sky Blue and the Midnight, have been developed by Mr E. Tresselt and put on the market by the Hunting Creek Fisheries. Breeding studies carried out by members of the GFSA, most notably Messrs A. Thomma and C. Perez, have shown that these are both true-breeding and genetically are homozygous transparent (i.e. matts). The Sky Blue is more or less a mid-blue colour while

Fig. 8.1 Spontaneous variants in fancy goldfish (see text).

the Midnight shows patches of solid black. Ideally perhaps these should be solid blue or black but although this is not often the case, they are attractive and interesting fish.

The most significant feature of these varieties is that they are homozygotes in which the suppression of three key features of the transparent gene's action has been achieved at least partially, namely inhibition of production and maintenance of the iridophores, melanophores and xanthophores. The suppression of these pigment cells in homozygotes is commonly almost complete but not necessarily so. Suppression of iridophores in run-of-the-mill matts produced in spawnings of the Shubunkin is virtually complete, that of the xanthophores is occasionally less than complete and matts with orange patches are not uncommon. Suppression of melanophores, while substantial, is very frequently incomplete. Matts often have a

Fig. 8.2 Hooded Telescope.

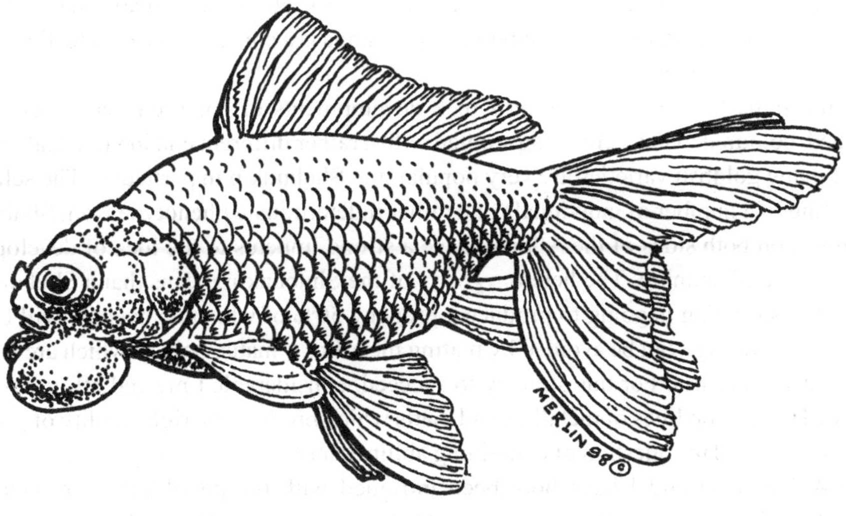

Fig. 8.3 Shi pao, a new Chinese variety characterised by water bubbles below the jaw.

peppering of fine black dots and an overall pale blue coloration. In the virtually complete absence of melanophores, body colour is pink due to haemoglobin present in the surface capillaries of the skin. The matt, non-reflective appearance is due to the absence of guanine backing on the scales and the absence of the subcutaneous layer of guanine-containing reflective tissue.

While the Shubunkin as a heterozygote consistently produces the typical 1 : 2 : 1 segregation of metallic, nacreous and matts, there are produced at a low frequency individuals which are difficult to assign unequivocally to a specific class. Usually these are intermediate between matts and nacreous; less frequently intermediates between metallics and nacreous are produced. The low frequency with which these problematical individuals are present does not significantly affect the segregation ratio but, on the principle that exceptions should be cherished, they repay further study and investigation in that they can give good indications as to what may be happening genetically that is unexpected but which indicates possible scope for genetic manipulation.

Among a number of unexpected results obtained from the Crucian carp × Shubunkin crosses was the production in F1 of a significant number of individuals which were in their phenotypes extreme matts or 'pinkies'. These showed virtually complete suppression of all three types of chromatophores. Their siblings were for the most part what one might have expected, mottled nacreous individuals. These tended to be predominantly orange in colour with dark spots and blotches and with production of hard gill covers and nacreous sheen. What was noticeable was the virtual absence of what one might call Shubunkin blue. After some years, areas clear of orange or brownish colour became suffused with a pale blue. This situation, though unusual and rather exceptional, shows, if further indication was needed, that dominance relationships are not fixed. In the pink heterozygous individuals the transparent mutant is behaving as a complete dominant, a demonstration that when the genetic background has not been fixed by virtue of the selection for the Shubunkin phenotype, for example, its segregation can produce wide phenotypic variation. It is arguable that the typical 1 : 2 : 1 segregations produced in Shubunkin spawnings is a consequence of selection for the nacreous, mottled phenotype.

At the same time as the Hunting Creek developments of coloured matts were being launched, coloured matts were being studied with a rather different end in view. Calico variants of many goldfish varieties are very popular, the Shubunkin in particular. The selection and testing of a number of individuals of a rather indeterminate phenotype from Shubunkin spawnings on both sides of the Atlantic gave a strong impetus to the idea of developing a true-breeding Shubunkin. The heterozygosity of the Shubunkin means that a Shubunkin × Shubunkin spawning is going to produce a progeny 50% of which on average will be discarded. This wastage can be avoided by mating metallics × matts, both of which are derived from Shubunkin lines. For this strategy to succeed, considerable time and effort has to be expended to develop lines of metallics and matts which produce the right quality of progeny when combined. This question of combining ability cannot be taken for granted.

Mr A. Thomma and I have both been intrigued with the possibility of producing a Shubunkin-like coloured matt homozygote. He has produced and tested homozygous individuals from American Shubunkin lines which have colour development and intensity equal to that of good Bristol Shubunkins. Their only deficiency was in the lack of the profuse dark spotting which is characteristic of the best Shubunkins. There would seem to be no good reason at present to suppose that this final hurdle could not be cleared. This situation gives a clear indication of how the application of genetic knowledge could improve the efficiency of production of young fish of satisfactory quality. Likewise, the same strategy could be employed in developing true-breeding Calico lines of any goldfish variety as long as sufficient motivation was present and an appropriate effort forthcoming.

While it is certainly understandable that breeders, both professional and amateur, want to carry out their operations most efficiently and make the best possible use of their time, energy, space and investment in their equipment, a very significant difference in attitude can emerge between them. The professional wants to maximise his profit whereas the amateur wishes to derive the maximum personal satisfaction from his hobby. Part of this satisfaction is derived from rising to and meeting a challenge. This may be a significant motive for many breeders. One breeder of my acquaintance stated publicly that he lost interest in breeding tropical fish because the offspring of any given pair were just like peas in a pod and he needed a challenge and for his own personal input to make a difference. Goldfish admirably met this need and he was satisfied with the level of pleasure he had derived over several decades. A committed goldfish breeder who produced in his day remarkable Calico Veiltails recounted his experience with a strain of Broadtail Moors which reproduced carbon copies of themselves. When he had established this fact he promptly lost interest and gave himself wholeheartedly to the Calico Veiltail.

Biotechnology and its actual and potential significance

Commercial interest in producing cheaply and effectively high quality fish in quantity has stimulated interest in gynogenesis (Cherfas 1975, Nagy *et al.* 1978, Nagy & Csanyi 1984). Gynogenesis is an asexual process in which unreduced eggs are produced which require activation by sperm which do not contribute to the genetic make-up of the progeny. The oocytes which normally produce eggs after undergoing meiosis, fail to complete the reduction division and mature with the somatic ($2n$) chromosome complement. Strains of goldfish and the Prussian carp show this mode of reproduction consistently and exist as clones. Gynogenesis can be induced by radiation treatment and so it is possible to induce it in fish which reproduce in the normal sexual fashion. It is possible, theoretically at any rate, to induce gynogenesis in selected fish, producing a maternal clone in which the process could be repeated, or which could be allowed to reproduce sexually if the treatment were not repeated.

This kind of manipulation comes under the general umbrella of 'biotechnology'. One aspect of this of current relevance is the use of sex hormones, which can be used in two distinct ways, one to initiate spawning at predetermined times, the second to control sexual development of fish. In goldfish the sex chromosome constitution of the female is *XX* and that of the male *XY*. By administration of the female sex hormones during development it is possible to produce *XY* females; with administration of male hormone extracts it is possible to produce *XX* males. Manipulation of sex is a useful tool which has been applied extensively in the production of farmed food fish such as rainbow trout, where the growth pattern of the female is commercially preferable to that of the male (which matures earlier than desired). It is possible that it could be useful in liberating the genetic potential of gynogenetic female clones were one to produce males artificially which might be capable of producing viable functional sperm.

At the present time (AD 2000) Biotechnology is highly controversial and use of sex-reversal techniques such as those just outlined is probably regarded with less favour now than it was say 20 years ago. Even more controversial is genetic engineering or genetic modification. Many at the present time are prepared to dismiss it out of hand while others are equally prepared to accept all it stands for uncritically. This polarisation has totally inhibited a rational discussion because at the extremes both sides are guilty of over-simplification. Each and every example of genetic modification is in a sense unique and requires to be evaluated in its own right. This is, however, a tedious and expensive business. One might well ask what relevance has this for the goldfish breeder. At the present time one can rightly say, very little. Genetic engineering is a very expensive business and it is highly unlikely that any capitalist (venture or otherwise) would be prepared to invest in projects to modify the goldfish genetically by the use of DNA technology. Perhaps the following brief discussion might help reassure those who feel that the use of this technology to modify goldfish might bring about significant improvement and to relieve any frustration that they might feel at being unable to follow this path.

Genetic modification has been attempted in two groups of food fish, the Salmonids and the Cichlids. The so-called 'anti-freeze' gene has been isolated from an arctic species, cloned and incorporated into the genome of species which have no such gene. The argument is that transgenic fish containing such a gene in a functional state would be able to grow and survive in an arctic environment. This argument is over-simplistic and encourages the false assumption that this one gene determines survival or otherwise of arctic salmonids within the Arctic Circle. Adaptation to such a harsh environment is unlikely to depend on a single gene but rather on a co-adapted complex of genes which functions in concert with the 'anti-freeze' gene itself. There is a similar over-simplification in the case of the induction of 'growth factor' transgenics in the Cichlid *Tilapia*. It was assumed that producing larger fish was a simple matter of increasing endogenous production of growth factor. Enhancement of growth rate and size must be achieved in a balanced way; a complex of genes is involved rather than a single gene. In genetic engineering it is frequently overlooked that the whole genotype must adapt to the presence of the introduced gene and a new co-adapted complex evolve. Such changes are unlikely to be instant fixes and the process of adaptation may be lengthy. Genes from closely related organisms will be more easily assimilated than those from distant relatives. The adjustment of the background genotype to the presence of an alien gene may be achieved easily or with difficulty.

It is unfortunate that in the present political and economic climate the 'quick fix' has an almost irresistible appeal. In presenting proposals for support of expensive research and genetic engineering, attention is focused on potential and the obvious pitfalls are either overlooked or assumed to be non-existent. Very often genetic engineers are biochemists rather than biologists and liable to be lacking in appreciation of biological facts of life. All too frequently they appear to be ignorant of the extent of their own ignorance and arrogant with it. The objective scientist these days, prepared to consider and not dismiss negative possibilities, is likely to receive very short shrift at the hands of the establishment apparatchiks. What truly needs to be borne in mind in that genes do not act in isolation, they also interact with other genes. These interactions may be quite unpredictable in both their nature and extent. A second aspect also overlooked is that there may be pleiotropic effects: there may be an obvious major gene effect but there may be others which can be overlooked. This is a possibility we ignore at our peril. Eventually we may hope that an objective appraisal of both benefits and hazards of the new technology will be achieved for the benefit of the human race.

One area where biotechnology can have positive and non-controversial application is in the study of goldfish evolution. It may well be possible by comparative biochemical studies of different goldfish varieties and wild relatives to fill in gaps in our knowledge. It could help to resolve some of the unsolved riddles such as the origin of the Oranda. Such a study would require the identification of systems which have generated sufficient measurable variation within a time frame of 1000 years. It is an unfortunate fact that the generation of morphological variation under domestication frequently outstrips that of detectable biochemical differences.

Population aspects

It seems reasonable to conclude that the improvement of the goldfish and its future will depend on what kind of a gene pool we can maintain and what steps we can take to conserve and augment it. If we consider gene pools we need to look to the populations of the goldfish and how they have probably changed over the past millennium.

The genetic population structure of wild and semi-domesticated populations of the goldfish can be considered usefully in terms of the Hardy Weinberg equilibrium concept. The latter is very robust and is capable of fine-tuning to accommodate situations which can arise when small samples are taken from a relatively large wild population to be the founders of quite small semi-domesticated populations. The major consequences will be firstly the 'founder effect': the sample taken, particularly if small, may not be truly representative of the original population. If this was to include individuals carrying xanthic mutant genes then the further consequences of small population size could have important repercussions. Genetic drift could result in chance changes of allelic frequency as a result of which, for example, xanthic allelic frequencies could either increase or decrease to the point of fixation or elimination. In small populations the coefficient of inbreeding (F) would increase which would lead to an increase in the proportion of homozygotes produced, which could of course include xanthic mutants.

If as might be expected the xanthic mutant was initially recessive the frequency of xanthic mutant phenotypes would be low, q^4 in a tetraploid rather than q^2 as in the diploid. However, if these individuals were preferentially selected for keeping, and not eaten, the xanthic mutation would then be at a selective advantage. Change in the dominance relationship between wild-type and mutant alleles might then come about through selection of the background genotype.

This process, initiated possibly in the T'ang dynasty, could have been sustained during the Song period when perhaps the change if not complete could have been well advanced. During the Ming period population sizes would have been even smaller and the level of inbreeding would have increased. This would have increased the selection pressure in changing dominance relations between the twin-tail allele and its wild-type counterpart. The profusion of variants which arose in the Qing period could very well have been a consequence of the very large number of scattered, relatively inbred populations. This population structure would maximise the opportunities for a wide range of mutations to occur and achieve homozygosity and thus be expressed.

Very strong selection pressures have undoubtedly been exerted by breeders in the case of both the xanthic and the twin-tail mutations. These pressures, especially in the case of the twin-tail example, have no doubt been opposed by natural selection pressures in favour of the normal phenotype. It is possible that incompletely penetrant dominance is the resultant of these two opposing selection pressures, dominance of the mutant being favoured by the breeder and a higher level of fitness (and the normal phenotype) by natural selection pressures. This opposition of selection pressures is more obvious in the case of the twin-tail than the xanthic mutation. Xanthism may be more nearly neutral in its selective value; in turbid waters the selective disadvantage would be less than in clear waters. The feeding habits of carp tend to generate turbidity which would certainly lessen the disadvantage of conspicuous coloration.

The importance of the protected aquarium environment in favouring diversification of fancy goldfish variants cannot be overestimated. The drawback of this is that it has tended to create more numerous, almost closed, very small populations, in danger of becoming inbred and inviable. One might well apply the critical mass concept to these small populations which of their very nature have small gene pools and are liable to inbreeding depression. The great imponderable in this situation is the frequency with which deleterious recessive are present. There will be a tendency for any of those present sooner or later to become homozygous and if this happens to a significant number of them the population could become poorly viable and pass the point of no return. This is of course the position which endangered species are in and from which in extreme cases there may be no hope of recovery. The closest parallel is with rare breeds which, when a critical stage has been reached and severe inbreeding depression is about to set in, may be revived by outcrossing and backcrossing. A practical remedy for this problem is an idea initiated and fostered by the Goldfish Society of America, that of the Breeders' Circle, in which groups of enthusiasts for particular varieties maintain collaboratively a much larger gene pool collectively than they would as individuals. This problem bears more heavily on the Occidental than the Oriental breeder.

When endangered species, rare breeds and varieties are considered, the would-be conservationist is almost bound to come up against notions of genetic purity which have semi-political overtones smacking of discredited eugenic theory. To some, interspecific or even intervarietal hybrids are anathema. This point of view is favoured by rigid adherence to the prescriptions of breed societies. These serve well enough when the available gene pool of the breed is large but are not in the interest of the breed's long-term survival when its size declines drastically. Breeds evolve, and the Hereford or Aberdeen Angus cattle of today are not what they were 50 or 100 years ago. To maintain varieties of goldfish in a satisfactory genetic state, carefully controlled inward gene flow or introgression may be necessary to revitalise a variety's flagging condition.

There is nothing unnatural in interspecific or other inter-population hybrids; in nature the vast majority will be weeded out by natural selection but those with advantageous features could well survive and multiply. Gene flow between some species of plants and animals appears to be an on-going phenomenon.

Envoi

What of the Future?

Genetics can be considered at several levels, the molecular, the cellular, the organism and the population, all of which are of interest and concern to those interested in goldfish improvement. At the molecular level, changes in pigmentation set the whole ball rolling in the T'ang dynasty. The chemistry of guanine is probably the best understood of all the compounds involved as it is one of the component bases of DNA. The pteridines and carotenoids have also been studied in depth, but there is room for further more effective communication between biochemists and enthusiasts who are aware of carotenoids and oblivious to pteridines. It would be good if some inducement could be found for more detailed categorisation of the xanthic pigments. Of the three recognisable colours red, orange and yellow, only the latter is unequivocally expressed. Whether a fish appears red or orange depends on the environmental conditions it has experienced. From Affleck's discussion in his 1952 paper the following hypothesis might well be worth investigation, that two distinct pigments are present, red and yellow, which are both present in the common orange-red individuals. Inhibition of either pigment could bring about significant colour change, that of the red would give rise to lemon or chrome yellow coloration while failure to produce yellow pigment would be expected to produce a stronger red colour. This could be the basis of the greatly admired but to date rather fugitive ox-blood colour. It may be possible to separate, identify and characterise these pigments, which would greatly simplify subsequent genetic analysis.

Kajishima's *Dp1/Dp2* loci control loss primarily of melanin followed by that of xanthic pigments in some instances. The blue mutant of Chen (1925) is significant in that it apparently determines loss of xanthophores without effect on melanophores. It would be useful to establish whether this mutation is implicated in the development of Magpie colour patterns and the absence of the bronze tint from the abdomens of Moors in which less intense black pigmentation is apparent. It would also be useful to clarify the situation regarding melanic pigments, whether or not distinct black and brown pigments are produced.

Compared with the metallic varieties, the issue of colour in calico varieties is perhaps even less well understood at the fundamental level. In the presence of the *Dp1* or *Dp2* alleles the effect of the transparent mutation appears to be straightforward; similar colour patterns can be produced, such as Sarassa, in both metallic and nacreous fish. In the absence of the demelanising effect the typical Calico mottling effect can be produced. As far as we can judge this is not an effect solely of the absence of guanine from the scales and the presence of both melanophores and xanthophores, since individual net-like transparent fish with little or no guanine deposition on the scales do not show this mottling. The presence of the transparent mutant and the wild-type pigment genes is not of itself a guarantee of mottling and development of typical calico patterns with development of the desirable blue ground colour. In the experimental Crucian carp × London Shubunkin crosses most of the F1 were orange and only after several years in relatively few individuals did anything approaching the standard calico pattern develop. This suggests that background genotype is an important determinant of calico pattern development.

At the cellular level and specifically at that of the chromosomes, study of their behaviour would be particularly informative. There is good evidence of polyploidy influencing inheritance patterns, notably the duplicate inheritance of the depigmentation mutant genes. The inheritance of the telescope eye is an interesting case of transition from the polyploid (tetrasomic) pattern in the cross Telescope × wild-type to the diploid (disomic) pattern shown in crosses between domesticated goldfish varieties. Is this an example of gene silencing and the transition from the buffered polyploid genetic system which is relatively unresponsive to selection pressures to the much more responsive diploid system? Where we have monofactorial inheritance is this a result of gene silencing and extensive diploidisation of the genome? In terms of its polyploidy, the goldfish seems to behave like a segmental allopolyploid with its combination of disomic and tetrasomic inheritance. Studies of the regularity or otherwise of meiotic chromosome behaviour could clarify the situation regarding the occasional production of aberrant progeny individuals which could be explained on the basis of cytogenetic accidents such as chromosome structural changes, position effects and aneuploidy.

At the level of the individual organism there is a lack of hard evidence regarding developmental outcomes of the action of many mutant genes. The hypotheses we have formed are based frequently on anecdotal evidence which is in many instances consistent and garnered from diverse sources and a wide geographic range. A programme of controlled experiments exploring environmental effects on development of the twin-tail in a range of varieties would be useful. The author has heard of an unusual pearlscale phenocopy induced by environmental shock during the course of embryonic development.

The future of exotic varieties of the goldfish depends very much on how successful we are in maintaining their gene pools, not so much in their obvious morphological distinctive characters as in the genes whose effects are difficult to evaluate but the balance between which determines the vigour and viability of populations. Where varieties are popular and widespread there is likely to be little problem, but varieties such as the Tosakin and Jikin which are difficult to manage could be at risk.

The fascination which the goldfish has had for both the fancier and scientists such as Chen, Matsui and Kajishima, has arisen both from the incredible array of diverse, often beautiful (and always interesting!) variants it has produced and the complexity of the evolutionary changes which have occurred. In the present age of expensive high-tech research, the goldfish is unlikely to attract appropriate investment in its own right, leaving ample scope for the well-informed amateur biologist to observe and contribute to knowledge. In astronomy there is a place for both the amateur observer as well as the highly sophisticated professional, with mutual respect between them. It is to be hoped that some similar relationship could arise between amateur and professional biologists in the cult of the goldfish.

Devotees undoubtedly feel, if they are of a scientific bent, considerable frustration at the lack of detailed and reliable information in a number of areas. The breeder, whether amateur and professional, usually cannot spare or make time to carry out any scientific investigation. On the other hand, the professional biologist engaged in scientific research is, as we have noted, unlikely to obtain support for the kind of study which is needed to consolidate the necessary genetic and cytogenetic basis for goldfish improvement. There does remain a potential resource which has not been tapped and that is of biologists in training at high schools, junior colleges and universities. Basic investigations of goldfish development, genetics, cytogenetics, physiology and behaviour could provide useful subjects for ongoing projects. These would provide a useful introduction for students intending to specialise in aquaculture, aquatic ecology and developmental biology.

Glossary

Additive gene action Cumulative effects on a particular aspect of the phenotype produced by alleles at the same or different gene loci.

Albinism Produced by loss of melanin. Goldfish albinos have pink eyes with body colour, red, yellow or white dependent on whether xanthic pigments are produced. Two gene loci control albinism P/p and C/c, ppcc are albino.

Allele (allelomorph) Alternative states of an hereditary Mendelian factor.

Allopolyploid Having two or more distinct genomes.

Allosomes The sex (X and Y) chromosomes.

Aneuploidy The addition to or loss of individual chromosomes from the genome.

Autopolyploid Having more than two homologous genomes.

Autosomes Chromosomes other than the X and Y chromosomes.

Azumanishiki Calico Oranda.

Background genotype effect Interactive effects between a specific locus and other unspecified genes in the genome.

Bivalents Pairs of homologous chromosomes formed in meiosis.

Chakin Chocolate, brown or purple goldfish.

Chiasma A cross-like configuration observed in the first division of meiosis (*see* Crossing-over).

Chin(chi)yu Common Goldfish.

Chi yu Wild Goldfish.

Chotengan Celestial.

Chromatophores The pigment cells containing melanin-melanophores; xanthic pigments-xanthophores or guanine-iridophores.

Chromosome The thread-like bodies within the cell nucleus which carry the genes.

Crossing-over The exchange of chromosome segments occurring in meiosis which produces recombination of genes.

Cytogenetics The cytological aspects of heredity.

Cytology The study of the cell, particularly of the chromosomes.

Deficiency or deletion Loss of a chromosome segment.

Deme-ranchu Celestial.

Depigmentation The process by which loss of chromatophores and the pigments they contain occurs. Controlled in goldfish by duplicate dominant genes Dp1 and Dp2.

Diploid Having two homologous sets of chromosomes (or genomes).

Diploidisation The establishment of a diploid inheritance pattern.

Disomic transmission The normal pattern of chromosome segregation produced by pairs of homologous chromosomes in meiosis.

Dominant The character of a contrasting pair of hereditary characters which is expressed in F1.

Duplication A repeated chromosome segment.

Epistasis Control of expression of one gene by another at a different locus, the latter is epistatic, the former hypostatic.

Euploidy The possession of complete chromosome sets, i.e. a balanced chromosome complement.

Expressivity The extent to which the mutant phenotype is developed.

Founder effect The significant divergence in gene frequencies produced between an original and derived population when the founder members of the latter are a small and genetically unrepresentative sample from the original.

Funa Wild goldfish.

Gene The unit of heredity which determines expression of a specific character.

Gene pool The total genetic resources of a population.

Genetic drift Changes in gene frequencies of small populations produced by random fluctuation.

Genetic engineering The deliberately planned and executed change in the genetic constitution of an organism.

Genetic modification The artificial change of the genetic constitution of an individual organism, usually involving the introduction of alien DNA.

Genetics The scientific study of variation and heredity.

Genome The complete but unduplicated gene complement of an organism – a complete set of chromosomes.

Genotype The hereditary constitution of the individual.

Hama tou Toadhead.

Hammerscales (or hammered scales) Scales with a concave surface, producing on reflection pin points of light, heredity not studied, anecdotal evidence suggest a recessive mutant.

Hanafusa Pompon with dorsal fin.

Haploid Having a single complete set of chromosomes (or one complete genome).

Heterogamety The production of two gametic types containing either the X or the Y chromosome.

Heterozygote An individual carrying two different alleles at the same locus.

Hibuna Common Goldfish.

Homogamety The production of gametes of a uniform type containing the X chromosome.

Homologous Having a common evolutionary origin.

Homozygote An individual carrying identical alleles at a specific locus.

Hypostasis The situation when the action of a gene at one locus is controlled by another at a different locus.

Interchange Exchange between non-homologous chromosomes of chromosome segments.

Inversion A reversed sequence of genes in part of a chromosome.

Ji-yu Wild Goldfish.

Jin yu Common Goldfish.

Kometto Comet.

Kujyaku Jikin (Peacock tail).

Kuro demekin Moor.

Linkage An association of parental characters in a progeny at a higher than expected frequency.

Locus The location of a gene in a chromosome.

Maruko Eggfish.

Matt Having no reflective tissue.

Meiosis The process of cell division which produces gametic nuclei in which the chromosome number is halved.

Metallic Having guanine-backed reflective scales.

Mitosis The process of cell division in which identical daughter cells are produced.

Monosomic A chromosome without a full homologue producing a haploid pattern of inheritance, e.g. the X and Y chromosomes in an XY individual.

Multivalent A group of three or more chromosomes formed in meiosis.

Mutation A sudden heritable change.

Nacreous Having transparent scales combined with reflective tissue beneath the scales.

Net-like transparent A recessive mutant reducing production of guanine in the scales producing a reticulate effect. Gene symbols N/n.

Oligogenes Genes of major effect.

Pearlscale A scale with a convex surface with a reflective boss produced in the central exposed portion of the scale. Probably due to a recessive mutation.

Penetrance (incomplete) This occurs when the expected phenotype fails to develop in an individual.

Phenocopy A phenotype mimicking that of a known mutant produced by an environmental shock.

Phenotype The actual expression of a character or characters in an organism arising from the genotype – environment interaction.

Phylogeny The evolutionary succession of biological variants.

Polygenes Genes of small and additive effect.

Polyploid Having more than two complete sets of homologous chromosomes.

Position effect The phenotypic change produced by the changed position of a gene in the genome.

Recessive The character of a contrasting pair of hereditary factors which is not expressed in F1 but which re-appears in F2.

Recombination The production of novel combinations of parental characters which arise in progeny of hybrids.

Rokurin Jikin (Peacock tail).

Sanshoku demekin Calico Telescope.

Seibun gyo Blue goldfish.

Sex linkage The pattern of inheritance of genes located in the non-homologous segments of the X and Y chromosomes.

Shukin Long-finned Lionhead.

Spindle The structure to which chromosomes become attached in cell division, which brings about separation of the chromosomes.

Suihogan Bubble-eye.

Tancho Redcap Oranda or Comet.

Telescope-eye Produced by a recessive mutation, gene symbols D/d. A single locus involved in crosses between domesticated goldfish. A bifactorial difference is observed in crosses with wild goldfish giving 15 : 1 segregation, possibly one duplicate locus is silenced in domesticated stock.

Test-cross A backcross to the homozygous recessive parent of a heterozygous individual.

Tetrasomic transmission The pattern of chromosome segregation produced when four homologous chromosomes take part in a meiotic division.

Tetsugyo Ironfish.

Tetsuwonaga Iron Fringetail.

Transformation The incorporation of alien genetic material in an organism.

Transparent mutant An incomplete dominant affecting production of all three chromatophore types. Gene symbol T/t.

Twin-tail Arises from duplication of the caudal fin fold in embryonic development with subsequent partial duplication of skeletal structures in the caudal peduncle. Duplication of caudal may be partial or compete. Produced probably by an incompletely penetrant dominant gene at a single locus.

Univalent An unpaired chromosome in meiosis.

Wild-type The normal condition found in an organism or at a gene locus.

Yamagata Twin-tailed Comet.

Zygote The product of the fusion of gametes in sexual reproduction.

Bibliography

Affleck, R.J. (1952) The nacreous (mottled) group of the goldfish (*Carassius auratus L.*) with an analysis of the colours seen in these fish. *Australian Journal of Marine and Freshwater Research,* **3**, 126–139.

Affleck, R.J. (1960) The embryology of the goldfish, *Carassius auratus L.* with notes on domesticated varieties. PhD thesis, Birkbeck College, University of London.

Allard, R.W. (1960) *Principles of Plant Breeding.* J. Wiley, New York and London.

Andrews, C. (1987) *Fancy Goldfishes*. Salamander Books, London.

Atkins, E.M. (1936) *Goldfish and other Cold-water Fishes*. Atkins, Sanderstead.

Bagnara, J.T., Matsumoto, J., Ferris, W., *et al.* (1979) Common origin of pigment cells. *Science,* **203**, 410–414.

Bateman, G.C. (1890) *Fresh-water Aquaria*. L. Upcott Gill, London.

Bateson, W. (1894) *Materials for the Study of Variation treated with Especial Regard to Discontinuity in the Origin of Species*. Macmillan & Co., London.

Betts, L.C. (1939) *The Goldfish*. Marshall Press, London.

Blanchon, H.L.A. (1912) *Le cyprin doré de la Chine et ses varieties*. Cosmos, Paris.

Carbonnier, P. (1872) Sur la reproduction et le dévelopment du poisson télescope. *Comptes rendues Hebdomadal Academie des Sciences (Paris),* 1127–1129.

Chen, S.C. (1925) Variation in external characteristics of goldfish *Carassius auratus. Contributions from the Biological Laboratory of the Science Society of China*, **1**, 1–64.

Chen, S.C. (1928) Transparency and mottling, a case of Mendelian inheritance in the goldfish, *Carassius auratus. Genetics,* **13**, 434–452.

Chen, S.C. (1934) The inheritance of blue and brown colours in the goldfish *Carassius auratus. Journal of Genetics,* **29**, 61–74.

Chen, S.C. (1956) A history of the domestication and the factors of the varietal formation of the common goldfish, *Carassius auratus. Scientia Sinica*, **5**, 287–321.

Cherfas, N.B. (1975) Investigation of radiation induced diploid gynogenesis in the carp (*Cyrinus carpio L.*). I. Experiments on obtaining the diploid gynogenetic progeny in mass quantities. *Genetika,* **11**, 78–86.

Darwin, C.R. (1868) *The variation of animals and plants under domestication*. John Murray, London.

Derham, A. (1926) *The Breeding of Fancy Goldfish.* Marshall's Printing Works, Harlow.

Evans, A. (1954) *Goldfish*. Muller, London.

Federation of British Aquatic Societies (FBAS) (1947) *Goldfish Standards*. FBAS.

Federation of British Aquatic Societies (FBAS) (1954) *Goldfish Standards*. FBAS.

Federation of British Aquatic Societies (FBAS) (1988) *Goldfish Standards*. FBAS.

Ferris, S.D. & Whitt, G.S. (1977) The evolution of duplicate gene expression in the carp (*Cyprinus carpio*) *Experientia,* **33**, 1299–1301.

Fisher, R.A. (1930) *The Genetical Theory of Natural Selection*. Oxford University Press, Oxford.

Fox, H.M. & Vevers, G. (1960) *The Nature of Animal Colours.* Sidgwick & Jackson, London.

Goldfish Society of America (GFSA) (1966) *Goldfish Breed Guidelines*. The Goldfish Report December 1996.

Goldfish Society of Great Britain (GSGB) (1972) *GSGB Standards.* GSGB.

Goldfish Society of Great Britain (GSGB) (1985) *GSGB Standards.* GSGB.

Goldfish Society of Great Britain (GSGB) (1995) *Nationwide Standards.* GSGB.

Goldfish Society of Great Britain (GSGB) *Varieties of Fancy Goldfish. Betts Memorial Booklets.* GSGB.
Goldfish Society of Great Britain (GSGB) *Colour. Betts Memorial Booklets.* GSGB.
Goldfish Society of Great Britain (GSGB) *Breeding, Rearing and Feeding Goldfish. Betts Memorial Booklets.* GSGB.
Hance, R.T. (1924) Heredity in Goldfish. *Journal of Heredity*, **15**, 177–182.
Hervey, G.F. (1950) *The Goldfish of China in the XVIII Century*. Sinological Series no. 3 China Society, London.
Hervey, G.F. & Hems, J. (1948) *The Goldfish*. Batchworth Press, London.
Hodge, A.E. & Derham, A. (1926) *Goldfish Culture for Amateurs*. Witherby, London.
Houghton, Rev. W. (1879) *British Freshwater Fishes*. Reprinted 1984 by Peerage Books, London.
Innes, W.T. (1917–1932) *Goldfish Varieties and Tropical Aquarium Fishes* (15 editions). Innes Publishing Co., Philadelphia.
Innes, W.T. (1947) *Goldfish Varieties and Water Gardens*. Innes Publishing Co., Philadelphia.
Kajishima, T. (1960a) Analysis of gene action in the transparent-scaled goldfish, *Carassius auratus*. I. On the gene action in the disappearance of guanophores, *Embryologia*, **5**, 107–126.
Kajishima, T. (1960b) The normal development stages of the goldfish, *Carassius auratus. Japanese Journal of Ichthyology*, **8**, 20–28.
Kajishima, T. (1977) Genetic and developmental analysis of some new colour mutants in the goldfish *Carassius auratus. Genetics*, **86**, 161–174.
Kajishima, T. & Takeuchi, I.K. (1977) Ultrastructural analysis of gene interaction and melanosome differentiation in the retinal pigment cells of the albino goldfish. *Journal of Experimental Zoology*, **200**, 349–358.
Kelsh, R.N., Brand, M., Yun-Jin Jiang *et al.* (1996) Zebrafish pigmentation mutations and the processes of neural crest development. *Development*, **123**, 369–389.
Kishinouye, K. (1898) Goldfish and other ornamental fish of Japan. *Natural Science*, **13**, 39.
Klug, W.S. & Cummings, M.R. (2000) *Concepts of Genetics*. Prentice Hall, Upper Saddle River, New Jersey.
Kobayasi, H., Kawashima, Y. & Takeuchi, N. (1970) Comparative chromosome studies in the genus *Carassius*, especially with a finding of polyploids in the Ginbuna (*C. auratus langsdorfii*). *Japanese Journal of Ichthyology*, **17**, 153–160.
Kobayasi, H., Nakano, K. & Nakamura, M. (1977) On the hybrids 4*n* ginbuna (*Carassius auratus langsdorfii*) × kinbuna (*Carassius subsp.*) and their chromosomes. *Bulletin of the Japanese Society of Scientific Fisheries*, **43**, 31–37.
Koh, T.P. (1931) Osteology of *Carassius auratus. Science Report National Ts'ing Hua University*, **1**: 61.
Koh, T.P. (1932) Osteological variations in the axial skeleton of goldfish (*Carassius auratus*). *Science Report National Ts'ing Hua University*, **2**, 109–121.
Koh, T.P. (1934) Notes on the evolution of goldfish. *China Journal of Science and Arts (Shanghai)*, **20**, 101–107.
Kuhn, F. *Der kleine Goldfisch teich*, Insel, Leipzig.
Li Zhen (1988) *Chinese Goldfish*. Foreign Languages Press, Beijing.
Lightcap, J. (2000) *GFSA Report*, vol. 28(1). Goldfish Society of America.
Liu, S., Sezaki, K., Hashimoto, K. & Nakamura, M. (1980) Distribution of polyploids of 'ginbuna' *Carassius auratus langsdorfii* in Japan. *Bulletin of the Japanese Society of Scientific Fisheries*, **46**, 413–418.
Man Shek-hay (1993) *Goldfish in Hong Kong*. Urban Council, Hong Kong.
Mather, K. (1973) *Genetical Structure of Populations*. Chapman & Hall, London.
Matsubara, S. (1908) Goldfish and their culture in Japan. *Bulletin of the United States Bureau of Fisheries*, **28**, 381–397.
Matsui, Y. (1930) Genetical studies on Fresh-water fish. 2. On the hybrids of *Cyprinus carpio* and *Carassius carassius (L.)*. *Journal of the Fisheries Experiment Station, Toyko*, no. 2.
Matsui, Y. (1933) Preliminary note on the inheritance of caudal and anal fins in goldfish of Japan. *Proceedings of the Imperial Academy, Tokyo*, **9**, 655–658.
Matsui, Y. (1934) Genetical studies on goldfish of Japan. *Journal of the Imperial Fisheries Institute*, **30**, 1–96.
Matsui, Y. (1971) *Goldfish*. Hoikusha, Osaka.
Matsui, Y. (1972) *Goldfish Guide*. Pet Library, Harrison, New Jersey.
Matsumoto, J., Kajishima, T. & Hama, T. (1960) Relation between the pigmentation and pterin derivatives of chromatophores during development in the normal black and transparent scaled type of goldfish (*Carassius auratus*). *Genetics*, **45**, 1177–1189.

Mettler, L.E. & Gregg, T.G. (1969) *Population Genetics and Evolution*. Prentice Hall, Englewood Cliffs, New Jersey.

Mills, D. (1985) *Keeping Goldfish*. Blandford Press Ltd., Poole.

Mullertt, H. (1883) *The Goldfish and its Systematic Culture*. Clarke, Cincinnati, Ohio.

Nagy, A. & Csanyi, V. (1984) A new breeding system using gynogenesis and sex reversal for fast inbreeding in carp. *Theoretical and Applied Genetics*, **67**, 485–490.

Nagy, A., Rajki, K., Horvath, L. & Csanyi, V. (1978) Investigation on carp *Cyprinus carpio L.* gynogenesis. *Journal of Fish Biology*, **13**, 215–225.

Newman, E. (1875) Telescope Fishes (*Carassius auratus*). *Zoologist* **2** ser. 10, 4501.

Ohno, S. (1970) *Evolution by Gene Duplication*, Springer-Verlag, Berlin.

Ojima, Y., Ueda, T. & Narikawa, T. (1979) A cytogenetic assessment on the origin of the goldfish. *Proceedings of the Japanese Academy, ser. B*, **55**, 58–63.

Onozato, H., Torisawa, M. & Kusama, M. (1983) Distribution of the gynogenetic polyploid crucian carp, *Carassius auratus* in Hokkaido, Japan. *Japanese Journal of Ichthyology*, **30**, 184–190.

Orme, F.W. (1979) *Fancy Goldfish Culture*. Saiga Publishing Co., Hindhead.

Pénzes, B. & Tölg, I. (1986) *Goldfish and Ornamental Carp*. Barron's Ltd., London.

Pereira, R.A. (1937) The Goldfish. *Hong Kong Naturalist*, **8**, 41–53.

Pouchet, G. (1870) Sur les cyprins monstrueux (*Cyprinus auratus*) venant de Chine. *Journal d'Anatomie et de la Physiologie, Paris*, **7**, 561–569.

Roughley, T.C. (1936) *The Cult of the Goldfish*. Angus & Robertson, Sydney.

Sauvigny de, B. (1780) *Histoire Naturelle des Dorades de la Chine*, Paris.

Schaeck de, M. (1893) Histoire du poisson doré (*Carassius auratus*). *Revue des Sciences Naturelles Appliquées* 1893, 111–120.

Schröder, J.H. (1976) *Genetics for Aquarists*. T.F.H. Publications, Neptune, New Jersey.

Schultz, R.J. (1980) Role of polyploidy in the evolution of fishes. In: *Polyploidy – Biological Relevance* (ed. W.H. Lewis). Plenum Press, New York.

Sharma, V.K. (1980) A peculiar case of abnormality of caudal fin in *Carassius auratus* (*Linn*). *Journal of the Inland Fisheries Society of India*, **12**, 127–128.

Simmonds, N.W. & Smartt, J. (1999) *Principles of Crop Improvement*. 2nd edn. Blackwell, Oxford.

Smartt, J. (1997) The future of the Shubunkin. *Aquarist and Pondkeeper*, **61** (12), 28–30.

Smartt, J. (1999) Goldfish × Crucian carp. *Aquarist and Pondkeeper*, **64** (5), 56–59.

Smartt, J. & Bundell, J.H. (1996) *Goldfish Breeding and Genetics*. T.F.H. Publications Inc., Neptune, New Jersey.

Smith, H.M. (1909) *Japanese Goldfish: Their Varieties and Cultivation*. W.F. Roberts Company, Washington.

Smith, H.M. (1924) Goldfish and their cultivation in America. *National Geographic*, October, pp. 375–400.

Sofradzija, A., Berberovi, L.J. & Hadziselimovic, R. (1978) Chromosome sets of *Carassius carassius* and *Carassius auratus gibelio*. *Acta Biologica Iugoslavica* (*E. Ichthyoly*), **10**, 135–143.

Spieler, R.E. (1971) A carp goldfish hybrid with no caudal fin. *Transactions of the Kansas Academy of Science*, **73**, 342–343.

Taylor, J.E. (1881) *The Aquarium*. David Bogue, London.

Teichfischer, B. (1994) *Goldfische in aller Welt*. Tetra Verlag, Melle Germany.

Teitler, N. (1981) *The T.F.H. Book of Goldfish*. T.F.H. Publications Inc., Neptune, New Jersey.

Tippitt, R.R. & Bennett, M.C. (1965) Rediscovery of the net-like transparent goldfishes. *The Aquarium*, December, pp. 10–11.

Vevers, G. (1982) *The Colours of Animals*, Studies in Biology no. 146. Edward Arnold, London.

Wang, C.Y. & Li, Y.L. (1982) Studies on the karyotype of goldfish (*Carassius auratus*). 1. A comparative study of the chromosomes in crucian and red dragon-eye goldfish. *Acta Genetica Sinica*, **9**, 238–242.

Wang, C. & Li, Y. (1983) Taxonomy and phylogeny of different varieties of the goldfish (*Carassius auratus*) in China. *Acta Zoologica Sinica*, **29**, 267–277.

Watanabe, Y. (1988) *Handbook of Goldfish*. Hello Publishing Company, Tokyo.

Watase, S. (1887) On the caudal and anal fins of goldfishes. *Journal of the Science College, Imperial University, Tokyo, Japan*, **1**, 247–267.

Weaver, R.F. & Hedrick, P.W. (1997) *Genetics*. Wm. C. Brown, Dubuque, Iowa.

Winfield, I.J. & Nelson, J.S. (1991) *Cyprinid Fishes – systematics, biology and exploitation*. Chapman & Hall,

London.

Wolf, H.T. (1908) *Goldfish Breeds*. Innes & Sons, Philadelphia.

Yamamoto, T. (1973) Inheritance of albinism in the goldfish, *Carassius auratus*. *Japanese Journal of Genetics*, **48**, 53–64.

Yamamoto, T. (1975) A YY male goldfish from mating estrone-induced XY female and normal male. *Journal of Heredity*, **66**, 2–4.

Yamamoto, T.O. (1977) Inheritance of nacreous-like scaleness in the Ginbuna, *Carassius auratus langsdorfii*. *Japanese Journal of Genetics*, **52**, 373–377.

Yamamoto, T. & Kajishima, T. (1968) Sex hormone induction of sex reversal in the goldfish and evidence for male heterogamety. *Journal of Experimental Zoology*, **168**, 215–222.

Zan, R. & Song, Z. (1980) Analysis and comparison between the karyotypes *of Cyprinus carpio* and *Carassius auratus* as well as *Aristichthys nobilis* and *Hypophthalmichthys molitrix*. *Acta Genetica Sinica*, **7**, 72–76.

Zhang Zhong-ge (1984) Goldfish. In: *Evolution of Domesticated Animals* (ed. I.L. Mason), pp. 381–385. Longman, London.

Index

Printed in the USA/Agawam, MA
September 17, 2026

910661.033